I0484291

FEDERAL EXECUTIVE TEAM

Director, Climate Change Science Program .... William J. Brennan

Director, Climate Change Science Program Office .... Peter A. Schultz

Lead Agency Principal Representative to CCSP;
Deputy Under Secretary of Commerce for Oceans and Atmosphere,
National Oceanic and Atmospheric Administration .... Mary M. Glackin

Product Lead, Geophysical Fluid Dynamics Laboratory,
National Oceanic and Atmospheric Administration .... Hiram Levy II

Synthesis and Assessment Product Advisory
Group Chair; Associate Director, EPA National
Center for Environmental Assessment .... Michael W. Slimak

Synthesis and Assessment Product Coordinator,
Climate Change Science Program Office .... Fabien J.G. Laurier

Special Advisor, National Oceanic
and Atmospheric Administration .... Chad A. McNutt

EDITORIAL AND PRODUCTION TEAM

Chair .... Hiram Levy II, NOAA/GFDL
Scientific Editor .... Jessica Blunden, STG, Inc.
Scientific Editor .... Anne M. Waple, STG, Inc.
Scientific Editor .... Christian Zamarra, STG, Inc.
Technical Advisor .... David J. Dokken, USGCRP
Graphic Design Lead .... Sara W. Veasey, NOAA
Graphic Design Co-Lead .... Deborah B. Riddle, NOAA
Designer .... Brandon Farrar, STG, Inc.
Designer .... Glenn M. Hyatt, NOAA
Designer .... Deborah Misch, STG, Inc.
Copy Editor .... Anne Markel, STG, Inc.
Copy Editor .... Lesley Morgan, STG, Inc.
Copy Editor .... Susan Osborne, STG, Inc.
Copy Editor .... Susanne Skok, STG, Inc.
Copy Editor .... Mara Sprain, STG, Inc.
Copy Editor .... Brooke Stewart, STG, Inc.
Technical Support .... Jesse Enloe, STG, Inc.

Climate Projections Based on Emissions Scenarios for Long-Lived and Short-Lived Radiatively Active Gases and Aerosols

Synthesis and Assessment Product 3.2
Report by the U.S. Climate Change Science Program
and the Subcommittee on Global Change Research

EDITED BY:
Hiram Levy II, Drew Shindell, Alice Gilliland,
Larry W. Horowitz, and M. Daniel Schwarzkopf

SCIENCE EDITOR: Anne M. Waple

August, 2008

Members of Congress:

On behalf of the National Science and Technology Council, the U.S. Climate Change Science Program (CCSP) is pleased to transmit to the President and the Congress this Synthesis and Assessment Product (SAP), *Climate Projections Based on Emissions Scenarios for Long-Lived and Short-Lived Radiatively Active Gases and Aerosols.* This is part of a series of 21 SAPs produced by the CCSP aimed at providing current assessments of climate change science to inform public debate, policy, and operational decisions. These reports are also intended to help the CCSP develop future program research priorities.

The CCSP's guiding vision is to provide the Nation and the global community with the science-based knowledge needed to manage the risks and capture the opportunities associated with climate and related environmental changes. The SAPs are important steps toward achieving that vision and help to translate the CCSP's extensive observational and research database into informational tools that directly address key questions being asked of the research community.

This SAP assesses climate projections based on emissions scenarios for long-lived and short-lived radiatively active gases and aerosols. It was developed with broad scientific input and in accordance with the Guidelines for Producing CCSP SAPs, the Information Quality Act (Section 515 of the Treasury and General Government Appropriations Act for Fiscal Year 2001 [Public Law 106-554]), and the guidelines issued by the Department of Commerce and the National Oceanic and Atmospheric Administration pursuant to Section 515.

We commend the report's authors for both the thorough nature of their work and their adherence to an inclusive review process.

Sincerely,

Carlos M. Gutierrez
Secretary of Commerce
Chair, Committee on Climate
Change Science and Technology
Integration

Samuel W. Bodman
Secretary of Energy
Vice Chair, Committee on Climate
Change Science and Technology
Integration

John H. Marburger
Director, Office of Science and
Technology Policy
Executive Director, Committee
on Climate Change Science and
Technology Integration

TABLE OF CONTENTS

Synopsis...V

Preface...IX

Executive Summary...1

CHAPTER

1 Introduction...7
Prologue...7
1.1 Historical Overview..8
1.2 Goals and Rationale..9
1.3 Limitations..10
1.4 Methodology...11
1.5 Terms and Definitions...11

2 15

Climate Projections from Well-Mixed Greenhouse Gas Stabilization
Emission Scenarios
Questions and Answers...15
2.1 Introduction...17
2.2 Well-Mixed Greenhouse Gas Emission Scenarios From SAP 2.1a............19
2.3 Simplified Global Climate Model (MAGICC)...................................20
2.4 Long-Lived Greenhouse Gas Concentrations and Radiative Forcings.....21
2.5 Short-Lived Gases and Particles and Total Radiative Forcing..............22
2.6 Surface Temperature: MAGICC AND IPCC Comparisons...................23
2.7 Climate Projections for SAP 2.1a Scenarios...................................23

3 27

Climate Change from Short-Lived Emissions Due to Human Activities
Questions and Answers...27
3.1 Introduction...28
3.2 Emissions Scenarios and Composition Model Descriptions.................29
  3.2.1 Emissions Scenarios 29
  3.2.2 Composition Models 31
  3.2.3 Tropospheric Burden...31
  3.2.4 Aerosol Optical Depth 34
3.3 Climate Studies...36
  3.3.1 Experimental Design..36
  3.3.2 Climate Models 36
  3.3.3 Radiative Forcing Calculations...37
  3.3.4 Climate Model Simulations 2000 to 2050 44
  3.3.5 Climate Simulations Extended to 2100 51

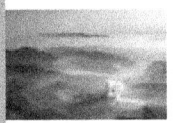

3.4 Regional Emission Sector Perturbations and Regional Models......................................53
   3.4.1 Introduction to Regional Emission Sector Studies 53
   3.4.2 Global Models...53
   3.4.3 Impact of Emission Sectors on Short-Lived Gases and Particles 53
   3.4.4 Regional Downscaling Climate Simulations...58

4 61

Findings, Issues, Opportunities, and Recommendations

4.1 Introduction 61
4.2 Key Findings 61
4.3 Issues Raised 62
   4.3.1 Emission Projections...62
   4.3.2 Particles (Indirect Effect, Direct Effect, Mixing, Water Uptake)...........................63
   4.3.3 Climate and Air Quality Policy Interdependence...64
4.4 Research Opportunities and Recommendations 65
   4.4.1 Emission Scenario Development..65
   4.4.2 Particle Studies (Direct Effect, Indirect Effect, Mixing, Water Uptake) 66
   4.4.3 Improvements in Transport, Deposition, and Chemistry 66
   4.4.4 Recommendations for Regional Downscaling 67
   4.4.5 Expanded Analysis and Sensitivity Studies..67
4.5 Conclusion 68

Appendix A 69

IPCC Fourth Assessment Report Climate Projections
(Supplemental to Chapter 2)

A.1 Mean Temperature...69
A.2 Temperature Extremes...69
A.3 Mean Precipitation 69
A.4 Precipitation Extremes and Droughts..70
A.5 Snow and Ice...70
A.6 Carbon Cycle..70
A.7 Ocean Acidification..70
A.8 Sea Level 70
A.9 Ocean Circulation 70
A.10 Monsoons 70
A.11 Tropical Cyclones (Hurricanes and Typhoons) 71
A.12 Midlatitude Storms 71
A.13 Radiative Forcing..71
A.14 Climate Change Commitment (Temperature and Sea Level)..71

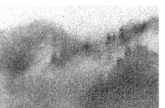

Appendix B 73
MAGICC Model Description
(Supplemental to Chapter 2)

Appendix C 75
Composition Models
(Supplemental to Chapter 3)
C.1 Geophysical Fluid Dynamics Laboratory 75
C.2 Goddard Institute for Space Studies 76
C.3 National Center for Atmospheric Research 77

Appendix D 79
Climate Models
(Supplemental to Chapter 3)
D.1 Geophysical Fluid Dynamics Laboratory 79
D.2 Goddard Institute for Space Studies 79
D.3 National Center for Atmospheric Research 80

Appendix E 81
Scenarios
(Supplemental to Chapter 4)
E.1 The Emissions Scenarios of the IPCC Special Report on Emission Scenarios 81
E.2 Radiative Forcing Stabilization levels and Approximate
    Carbon Dioxide Concentrations from the CCSP SAP 2.1a Scenarios 82

Glossary and Acronyms 83

References 85

Preface   **Authors:** Hiram Levy II, NOAA/GFDL; Drew T. Shindell, NASA/GISS; Alice Gilliland, NOAA/ARL; M. Daniel Schwarzkopf, NOAA/GFDL; Larry W. Horowitz, NOAA/GFDL
Contributing Author: Anne Waple, STG Inc.

Executive Summary   **Lead Authors:** Hiram Levy II, NOAA/GFDL; Drew T. Shindell, NASA/GISS; Alice Gilliland, NOAA/ARL; M. Daniel Schwarzkopf, NOAA/GFDL; Larry W. Horowitz, NOAA/GFDL
Contributing Author: Anne Waple, STG Inc.

Chapter 1   **Lead Authors:** Hiram Levy II, NOAA/GFDL; Drew T. Shindell, NASA/GISS; Alice Gilliland, NOAA/ARL, M. Daniel Schwarzkopf, NOAA/GFDL; Larry W. Horowitz, NOAA/GFDL
Contributing Authors: Anne Waple, STG Inc.; Ronald J Stouffer, NOAA/GFDL

Chapter 2   **Lead Authors:** Hiram Levy II, NOAA/GFDL; Drew T. Shindell, NASA/GISS
Contributing Author: Tom Wigley, NCAR

Chapter 3   **Lead Authors:** Drew T. Shindell, NASA/GISS; Hiram Levy II, NOAA/GFDL; Alice Gilliland, NOAA/ARL; M. Daniel Schwarzkopf, NOAA/GFDL; Larry W. Horowitz, NOAA/GFDL
Contributing Author: Jean-Francois Lamarque, NCAR; Yi Ming, UCAR

Chapter 4   **Lead Authors:** Hiram Levy II, NOAA/GFDL; Drew T. Shindell, NASA/GISS; Alice Gilliland, NOAA/ARL

Appendix A   **Lead Author:** Hiram Levy II, NOAA/GFDL
Contributing Author: Tom Wigley, NCAR

Appendix B   **Lead Author:** Hiram Levy II, NOAA/GFDL
Contributing Author: Tom Wigley, NCAR

Appendix C   **Lead Authors:** Larry W. Horowitz, NOAA/GFDL; M. Daniel Schwarzkopf, NOAA/GFDL; Drew T. Shindell, NASA/GISS
Contributing Author: Jean-Francois Lamarque, NCAR; Tom Wigley, NCAR

Appendix D   **Lead Authors:** Drew T. Shindell, NASA/GISS; M. Daniel Schwarzkopf, NOAA/GFDL; Hiram Levy II, NOAA/GFDL; Larry W. Horowitz, NOAA/GFDL
Contributing Author: Jean-Francois Lamarque, NCAR

Appendix E   **Lead Authors:** Drew T. Shindell, NASA/GISS; Hiram Levy II, NOAA/GFDL
Contributing Author: Anne Waple, STG Inc.

The influence of greenhouse gases and particle pollution on our present and future climate has been widely examined and most recently reported in the Intergovernmental Panel on Climate Change (IPCC) Fourth Assessment Report. While both long-lived (e.g., carbon dioxide) and short-lived (e.g., soot) gases and particles affect the climate, previous projections of future climate, such as the IPCC reports, have focused largely on the long-lived gases. This U.S. Climate Change Science Program (CCSP) Synthesis and Assessment Product provides a different emphasis.

We first examine the effect of long-lived greenhouse gases on the global climate based on updated emissions scenarios produced by another CCSP Synthesis and Assessment Product (SAP 2.1a). In these scenarios, atmospheric concentrations of the long-lived greenhouse gases leveled off, or stabilized, at predetermined levels by the end of the twenty-first century (unlike in the IPCC scenarios). However, the projected future temperature changes, based on these stabilization emissions scenarios, fall within the same range as those projected for the latest IPCC report. We confirm the robust future warming signature and other associated changes in the climate.

We next explicitly assess the effects of short-lived gases and particles. Their influence is found to be global in nature, substantial when compared with long-lived greenhouse gases, and potentially extending to the end of this century. They can significantly change the regional surface temperature, and by the year 2100 short-lived gases and particles may account for as much as 40 percent of the warming over the summertime continental United States. It is noteworthy that the simulated climate response to these pollutants is not confined to the geographical area where they are released. This implies a strong linkage between regional air quality control strategies and global climate change. We identify specific emissions reductions that would lead to benefits for both air quality and climate change mitigation, including North American surface transportation and Asian domestic fuel burning. The results reveal the necessity for explicit and consistent inclusion of the short-lived pollutants in assessments of future climate.

RECOMMENDED CITATIONS

For the Report as a whole:

CCSP, 2008: *Climate Projections Based on Emissions Scenarios for Long-Lived and Short-Lived Radiatively Active Gases and Aerosols*. A Report by the U.S. Climate Change Science Program and the Subcommittee on Global Change Research. H. Levy II, D.T. Shindell, A. Gilliland, M.D. Schwarzkopf, L.W. Horowitz, (eds.). Department of Commerce, NOAA's National Climatic Data Center, Washington, D.C., USA, 100 pp.

For the Preface:

Levy II, H., D.T. Shindell, A. Gilliland, M.D. Schwarzkopf, L.W. Horowitz, 2008: Preface in *Climate Projections Based on Emissions Scenarios for Long-Lived and Short-Lived Radiatively Active Gases and Aerosols*. H. Levy II, D.T. Shindell, A. Gilliland, M.D. Schwarzkopf, L.W. Horowitz, (eds.). A Report by the U.S. Climate Change Science Program and the Subcommittee on Global Change Research, Washington, D.C.

For the Executive Summary:

Levy II, H., D.T. Shindell, A. Gilliland, M.D. Schwarzkopf, L.W. Horowitz, 2008: Executive Summary in *Climate Projections Based on Emissions Scenarios for Long-Lived and Short-Lived Radiatively Active Gases and Aerosols*. H. Levy II, D.T. Shindell, A. Gilliland, M.D. Schwarzkopf, L.W. Horowitz, (eds.). A Report by the U.S. Climate Change Science Program and the Subcommittee on Global Change Research, Washington, D.C.

For Chapter 1:

Levy II, H., D.T. Shindell, A. Gilliland, M.D. Schwarzkopf, L.W. Horowitz, 2008: Introduction in *Climate Projections Based on Emissions Scenarios for Long-Lived and Short-Lived Radiatively Active Gases and Aerosols*. H. Levy II, D.T. Shindell, A. Gilliland, M.D. Schwarzkopf, L.W. Horowitz, (eds.). A Report by the U.S. Climate Change Science Program and the Subcommittee on Global Change Research, Washington, D.C.

For Chapter 2:

Levy II, H., D. T. Shindell, T. Wigley, 2008: Climate Projections From Well-Mixed Greenhouse Gas Stabilization Scenarios in *Climate Projections Based on Emissions Scenarios for Long-Lived and Short-Lived Radiatively Active Gases and Aerosols*. H. Levy II, D.T. Shindell, A. Gilliland, M.D. Schwarzkopf, L.W. Horowitz, (eds.). A Report by the U.S. Climate Change Science Program and the Subcommittee on Global Change Research, Washington, D.C.

For Chapter 3:

Shindell, D.T., H. Levy II, A. Gilliland, M.D. Schwarzkopf, L.W. Horowitz, 2008: Climate Change From Short-Lived Emissions Due to Human Activities in *Climate Projections Based on Emissions Scenarios for Long-Lived and Short-Lived Radiatively Active Gases and Aerosols*. H. Levy II, D.T. Shindell, A. Gilliland, M.D. Schwarzkopf, L.W. Horowitz, (eds.). A Report by the U.S. Climate Change Science Program and the Subcommittee on Global Change Research, Washington, D.C.

For Chapter 4:

Levy II, H., D.T. Shindell, A. Gilliland, 2008: Findings, Issues, Opportunities, and Recommendations in *Climate Projections Based on Emissions Scenarios for Long-Lived and Short-Lived Radiatively Active Gases and Aerosols*. H. Levy II, D.T. Shindell, A. Gilliland, M.D. Schwarzkopf, L.W. Horowitz, (eds.). A Report by the U.S. Climate Change Science Program and the Subcommittee on Global Change Research, Washington, D.C.

For Appendix A:

Levy II, H., T. Wigley, 2008: IPCC 4th Assessment Climate Projections in *Climate Projections Based on Emissions Scenarios for Long-Lived and Short-Lived Radiatively Active Gases and Aerosols*. H. Levy II, D.T. Shindell, A. Gilliland, M.D. Schwarzkopf, L.W. Horowitz, (eds.). A Report by the U.S. Climate Change Science Program and the Subcommittee on Global Change Research, Washington, D.C.

For Appendix B:

Levy II, H., T. Wigley, 2008: MAGICC Model Description in *Climate Projections Based on Emissions Scenarios for Long-Lived and Short-Lived Radiatively Active Gases and Aerosols*. H. Levy II, D.T. Shindell, A. Gilliland, M.D. Schwarzkopf, L.W. Horowitz, (eds.). A Report by the U.S. Climate Change Science Program and the Subcommittee on Global Change Research, Washington, D.C.

For Appendix C:

Shindell, D.T., L.W. Horowitz, M.D. Schwarzkopf, 2008: Composition Models in *Climate Projections Based on Emissions Scenarios for Long-Lived and Short-Lived Radiatively Active Gases and Aerosols*. H. Levy II, D.T. Shindell, A. Gilliland, M.D. Schwarzkopf, L.W. Horowitz, (eds.). A Report by the U.S. Climate Change Science Program and the Subcommittee on Global Change Research, Washington, D.C.

For Appendix D:

Shindell, D.T., M.D. Schwarzkopf, H. Levy II, L.W. Horowitz, 2008: Climate Models in *Climate Projections Based on Emissions Scenarios for Long-Lived and Short-Lived Radiatively Active Gases and Aerosols*. H. Levy II, D.T. Shindell, A. Gilliland, M.D. Schwarzkopf, L.W. Horowitz, (eds.). A Report by the U.S. Climate Change Science Program and the Subcommittee on Global Change Research, Washington, D.C.

For Appendix E:

Levy II, H., D.T. Shindell, 2008: Scenarios in *Climate Projections Based on Emissions Scenarios for Long-Lived and Short-Lived Radiatively Active Gases and Aerosols*. H. Levy II, D.T. Shindell, A. Gilliland, M.D. Schwarzkopf, L.W. Horowitz, (eds.). A Report by the U.S. Climate Change Science Program and the Subcommittee on Global Change Research, Washington, D.C.

PREFACE

Report Motivation and Guidance for Using this Synthesis and Assessment Product

Authors: Hiram Levy II, NOAA/GFDL; Drew T. Shindell, NASA/GISS; Alice Gilliland, NOAA/ARL; M. Daniel Schwarzkopf, NOAA/GFDL; Larry W. Horowitz, NOAA/GFDL

Contributing Author: Anne Waple, STG Inc.

INTRODUCTION

The U.S. Climate Change Science Program (CCSP) was established in 2002 to coordinate climate and global change research conducted in the United States. Building upon and incorporating the U.S. Global Change Research Program of the previous decade, the program integrates federal research on climate and global change, as sponsored by 13 federal agencies and overseen by the Office of Science and Technology Policy, the Council on Environmental Quality, the National Economic Council, and the Office of Management and Budget.

A primary objective of the U. S. CCSP is to provide the best possible scientific information to support public discussion and government and private sector decision making on key climate-related issues. To help meet this objective, the CCSP has identified an initial set of 21 synthesis and assessment products that address its highest priority research, observation, and decision-support needs.

The CCSP is conducting 21 such activities, covering topics such as the North American carbon budget and implications for the global carbon cycle, coastal elevation and sensitivity to sea-level rise, trends in emissions of ozone-depleting substances and ozone recovery and implications for ultraviolet radiation exposure, and use of observational and model data in decision support and decision making. The stated purpose for this report, Synthesis and Assessment Product (SAP) 3.2, is to provide information to those who use climate model outputs to assess the potential effects of human activities on climate, air quality, and ecosystem behavior.

In an examination of the U.S. CCSP Strategic Plan, the National Research Council (NRC) recommended that synthesis and assessment products should be produced, with independent oversight and review from the wider scientific and stakeholder communities. To meet this goal, NOAA requested an independent review of SAP 3.2 by the NRC. The NRC appointed an *ad hoc* committee composed of eight members who provided their review findings, and recommendations, suggestions, and options for the authors to consider in revising the first draft of SAP 3.2. The revised second draft was then posted for public comment for 45 days. This third draft is in response to those public comments.

BACKGROUND AND GOALS

The initial mandate for Synthesis and Assessment Product 3.2 (SAP 3.2), which is still listed on the official CCSP website <http://www.climatescience.gov/Library/sap/sap-summary.php>, was to provide "Climate Projections for Research and Assessment Based on Emissions Scenarios Developed Through the Climate Change Technology Program". With the development of long-lived greenhouse gas scenarios by another Synthesis and Assessment Product (SAP 2.1a; Clarke *et al.*, 2007), our mandate evolved to "Climate Projections for SAP 2.1a Emissions Scenarios of Greenhouse Gases". These emissions scenarios[1] were for the long-lived[2], and therefore globally well-mixed, radiatively active gases (greenhouse gases), and were constrained by the requirement that carbon dioxide concentrations stabilize within 100 to 200 years at specified levels of roughly 450, 550, 650 and 750 parts per million (ppm). See Box P.1 for additional details.

---

[1] Emissions scenarios represent future emissions based on a coherent and internally consistent set of assumptions about the driving forces (*e.g.*, population change, socioeconomic development, and technological change) and their key relationships.

[2] Long-lived radiatively active gases of interest have atmospheric lifetimes that range from ten years for methane to more than 100 years for nitrous oxide. While carbon dioxide's lifetime is more complex, we think of it as being more than 100 years in the climate system. Due to their long atmospheric lifetime, they are well-mixed and evenly distributed throughout the lower atmosphere. Global atmospheric lifetime is the mass of a gas or an particle in the atmosphere divided by the mass that is removed from the atmosphere each year.

BOX P.1: Stabilization Emission Scenarios and Background From CCSP SAP 2.1a

Synthesis and Assessment Product (SAP) 2.1 (Clarke *et al.*, 2007) is an important precursor to this Product. It explores different scenarios that lead to greenhouse gas concentrations stabilizing at different (higher) levels in the future. Scenario analysis is a widely used tool for decision making in complex and uncertain situations. Scenarios are "what ifs"—sketches of future conditions (or alternative sets of future conditions) used as inputs to exercises of decision making or analysis. Scenarios have been applied extensively in the climate change context. Examples include greenhouse gas emissions scenarios, climate scenarios, and technology scenarios.

The scenarios in SAP 2.1a are called "stabilization emissions scenarios" because they are constrained so that the atmospheric concentrations of the long-lived greenhouse gases level off, or stabilize, at predetermined levels by the end of the twenty-first century. They explicitly treat the economic and technological drivers needed to generate each level of greenhouse gases. Further discussion is found in Box 1.2 of Chapter 1.

Preindustrial levels of carbon dioxide were approximately 280 parts per million (ppm), and are currently around 380 ppm—a third higher than prior to the industrial era and higher than at any other time in at least the last 420,000 years (CCSP SAP 2.2). The four stabilization levels for SAP 2.1a were constructed so that the carbon dioxide concentrations resulting from stabilization are roughly 450, 550, 650, and 750 ppm. While the Intergovernmental Panel on Climate Change (IPCC) has also examined greenhouse gas emission scenarios, and those provided by SAP 2.1a are generally within the envelope of the IPCC scenarios, SAP 2.1a is an alternative approach to developing a consistent set of long-lived greenhouse gas concentrations.

This Product (SAP 3.2) explores the climate implications of such greenhouse gas "stabilization emissions scenarios" via several different computer simulations. The results of these projections are presented in Chapter 2 of this Product.

The SAP 2.1a scenarios (Clarke *et al.*, 2007) did not explicitly address the direct influence of short-lived[3] drivers of climate: carbon and sulfate particles and lower atmospheric ozone. Therefore, we expanded our mandate to include "Short-Lived Radiatively Active[4] Gases and Aerosols[5]". These short-lived gases and aerosols (particles) are largely of human-caused origin, are important contributors to large-scale changes in atmospheric temperature and climate in general, and are primarily controlled for reasons of local and regional air quality. Therefore, this added portion of the report is a critical first step in examining the climate impact of future actions taken to reduce air pollution.

The Prospectus for Synthesis and Assessment Product 3.2 contained two charges to the authors of this Product:

1. Develop climate projections for a series of scenarios for long-lived greenhouse gases provided by Synthesis and Assessment Product 2.1a, "Scenarios of Greenhouse Gas Emissions and Atmospheric Concentrations and Review of Integrated Scenario Development and Application".
2. Investigate the contributions of four short-lived pollutants in the lower atmosphere: ozone and three types of particles (soot/elemental carbon, organic carbon, and sulfate), usually identified in scientific terms as aerosols[5].

Short-lived greenhouse gases and particles have received less attention than long-lived greenhouse gases in previous international assessments and were not explicitly treated in Synthesis and Assessment Product 2.1a (Clarke *et al.*, 2007) but, as this report describes, they may affect the future climate in a substantial manner. Although sources of these pollutants tend to be localized, their impact is felt globally. This is of direct relevance to policy decisions regarding pollution, air quality, and climate change.

[3] Short-lived radiatively active gases and particles of interest in the lower atmosphere have lifetimes of about a day for nitrogen oxides, a day to a week for most particles, and a week to a month for ozone. Their concentrations are highly variable and concentrated in the lowest part of the atmosphere, primarily near their sources.
[4] Radiatively active gases and particles absorb, scatter, and re-emit energy, thus changing the temperature of the atmosphere.
[5] Aerosols are very small airborne solid or liquid particles that reside in the atmosphere for at least several hours, with the smallest remaining airborne for days.

READER'S GUIDE TO SYNTHESIS AND ASSESSMENT PRODUCT 3.2

This Product includes an Executive Summary and four Chapters.

The **Executive Summary** presents the key results and findings, and recommends four critical areas of future research. It is written in non-technical language and is intended to be accessible to all audiences.

Chapter 1 provides an Introduction to this study, and is intended to provide all audiences with a general overview. It is written in non-technical language, which should be accessible to all readers with an interest in climate change. It includes background material, discusses the scope of and motivation for this study, addresses its goals and objectives, and identifies the issues that are not addressed. It also contains two Boxes, one providing non-technical definitions of important terms, and the other containing a clear and concise description of the computer models employed in this study.

Chapter 2 focuses on the long-lived greenhouse gases and a set of scenarios provided by Synthesis and Assessment Product 2.1a. The Statement of Findings and the Introductory Section 2.1 are written in non-technical language and are intended for the general reader. The remainder of Chapter 2 provides detailed technical information about specific computer models, the resulting climate simulations, and a detailed interpretation of the results. It is intended primarily for the scientific community.

A simplified global climate model, MAGICC[6], is used to simulate globally-averaged surface temperature increases for the stabilization emission scenarios, and the results are assessed in the context of the Fourth Assessment Report of the Intergovernmental Panel on Climate Change (IPCC) Working Group 1 (IPCC, 2007). These comparisons are used to answer the first four questions posed in our Prospectus:

Q1. Do SAP 2.1a emissions scenarios differ significantly from IPCC emission scenarios?

Q2. If the SAP 2.1a emissions scenarios do fall within the envelope of emissions scenarios previously considered by the IPCC, can the existing IPCC climate simulations be used to estimate 50- to 100-year climate responses for the CCSP 2.1a carbon dioxide emission scenarios?

Q3. What would be the changes to the climate system under the scenarios being put forward by SAP 2.1a?

Q4. For the next 50 to 100 years, can the climate projections using the emissions from SAP 2.1a be distinguished from one another or from the scenarios recently studied by the IPCC?

Chapter 3 attempts to assess the direction, magnitude, and duration of future changes in climate due to changing levels of short-lived radiatively active gases and particles of human-caused origin. This is an area of research that is still at the initial stages of exploration and which the IPCC Fourth Assessment Report, as well as previous IPCC reports, investigated only superficially.

First, the stabilization emissions scenarios and models used to generate them are discussed. Next, the chemical composition models[7] used to produce the global distributions of short-lived gases and particles that help to drive the climate models are introduced. Twenty-first century climate is then simulated with three state-of-the-art comprehensive climate models[8], and the results are then used to address the four questions raised in the second section of our Prospectus:

Q5. What are the impacts of the radiatively active short-lived gases and particles not explicitly the subject of SAP 2.1a?

Q6. How do the impacts of short-lived species (gases and particles) compare with those of the well-mixed greenhouse gases as a function of the time horizon examined?

Q7. How do the regional impacts of short-lived species (gases and particles) compare with those of long-lived gases in or near polluted areas?

Q8. What might be the climate impacts of mitigation actions taken to reduce the atmospheric levels of short-lived species (gases and particles) to address air quality issues?

The Statement of Findings and the Introductory Section 3.1 are written in non-technical language and are intended for the general reader. The remainder of the chapter provides detailed technical information about the models, the result-

---

[6] MAGICC is a two-component numerical model consisting of a highly simplified representation of a climate model coupled with an equally simplified representation of the atmospheric composition of radiatively active gases and particles. This model is adjusted, based on the results of more complex climate models, to make representative predictions of global mean surface temperature and sea-level rise.

[7] Chemical composition models are state-of-the-art numerical models that use the emission of gases and particles as inputs and simulate their chemical interactions, global transport by the winds, and removal by rain, snow, and deposition to the earth's surface. The resulting outputs are global three-dimensional distributions of the initial gases and particles and their products.

[8] Comprehensive climate models are a numerical representation of the climate based on the physical properties of its components, their interactions, and feedback processes. Coupled Atmosphere-Ocean (-sea ice) General Circulation Models (AOGCMs) represent our current state-of-the-art.

ing climate simulations, and our interpretation of the results. It is intended primarily for the scientific community.

Chapter 4 provides a summary of the key findings, identifies a number of scientific issues and questions that arise from our study, and identifies new opportunities for future research. The five most critical areas identified by this study are:

1. The projection of future human-caused emissions for the short-lived gases and particles;
2. The indirect and direct effects of particles and mixing between particle types;
3. Transport, deposition, and chemistry of the short-lived gases and particles;
4. Regional climate forcing *vs.* regional climate response;
5. Sensitivity studies of climate responses to short-lived gases and particles.

We have written Chapter 4, as much as is possible, in non-technical language, and it is intended for all audiences.

EXECUTIVE SUMMARY

Lead Authors: Hiram Levy II, NOAA/GFDL; Drew T. Shindell,
NASA/GISS; Alice Gilliland, NOAA/ARL; M. Daniel Schwarzkopf,
NOAA/GFDL; Larry W. Horowitz, NOAA/GFDL

Contributing Author: Anne Waple, STG Inc.

SYNOPSIS

The influence of greenhouse gases and particle pollution on our present and future climate has been
widely examined and most recently reported in the
Intergovernmental Panel on Climate Change (IPCC)
Fourth Assessment Report. While both long-lived[1]
(e.g., carbon dioxide) and short-lived[2] (e.g., soot) gases and particles affect the climate, previous
projections of future climate, such as the IPCC reports, have focused largely on the long-lived
gases. This U.S. Climate Change Science Program Synthesis and Assessment Product provides
a different emphasis.

We first examine the effect of long-lived greenhouse gases on the global climate based on updated
emissions scenarios produced by another CCSP Synthesis and Assessment Product (SAP 2.1a).
In these scenarios, atmospheric concentrations of the long-lived greenhouse gases leveled off,
or stabilized, at predetermined levels by the end of the twenty-first century (unlike in the IPCC
scenarios). However, the projected future temperature changes, based on these stabilization
emissions scenarios, fall within the same range as those projected for the latest IPCC report.
We confirm the robust future warming signature and other associated changes in the climate.

We next explicitly assess the effects of short-lived gases and particles. Their influence is found
to be global in nature, substantial when compared with long-lived greenhouse gases, and potentially extending to the end of this century. They can significantly change the regional surface
temperature, and by the year 2100 short-lived gases and particles may account for as much as
40 percent of the warming over the summertime continental United States. It is noteworthy
that the simulated climate response to these pollutants is not confined to the geographical
area where they are released. This implies a strong linkage between regional air quality control
strategies and global climate change. We identify specific emissions reductions that would lead
to benefits for both air quality and climate change mitigation, including North American surface
transportation and Asian domestic fuel burning. The results reveal the necessity for explicit and
consistent inclusion of the short-lived pollutants in assessments of future climate.

---

[1] Atmospheric lifetimes for the long-lived radiatively active gases of interest range from ten years for methane
to more than 100 years for nitrous oxide. While carbon dioxide's lifetime is more complex, we can think of
it as being more than 100 years in the climate system. As a result of their long atmospheric lifetimes, they are
well-mixed and evenly distributed throughout the lower atmosphere. Global atmospheric lifetime is the mass
of a gas or a particle in the atmosphere divided by the mass that is removed from the atmosphere each year.

[2] Atmospheric lifetimes for the short-lived radiatively active gases and particles of interest in the lower atmosphere are about a day for nitrogen oxides, a day to a week for most particles, and a week to a month for
ozone. As a result of their short lifetimes, their concentrations are highly variable in space and time and they
are concentrated in the lowest part of the atmosphere, primarily near their sources.

ES.1 KEY FINDINGS

These results constitute important improvements in our understanding of the influence of both long-lived gases and short-lived gases and particles. The Fourth Assessment Report of the IPCC recognized that most of the global-scale warming since the middle of last century was very likely due to the increase in greenhouse gas concentrations, and also that the warming has been partially damped by increasing levels of short-lived particle pollutants. However, while the IPCC models were coordinated in their use of greenhouse gas emissions scenarios, the short-lived pollutants were widely varying in the emissions scenarios used, and their future impacts were not isolated from those of the long-lived gases.

Changes in pollutant levels, primarily over Asia, may significantly increase surface temperature and reduce rainfall over the summertime continental United States.

This Synthesis and Assessment Product provides a more comprehensive and updated assessment of the relative future contributions of long and short-lived gases and particles, with special, explicit focus on the short-lived component. This study encompasses a realistic time frame over which available technological solutions can be employed, and this study, in particular, focuses on those gases and particles whose future atmospheric levels are also subject to reduction due to air pollution control.

1. Our results suggest that changes in short-lived gases and particles (pollutants) may significantly influence the climate, in the twenty-first century. By 2050, projected changes in short-lived pollutant concentrations in two of the three studies are responsible for approximately 20 percent of the simulated global-mean annual average warming (see Section 3.3.4 and Figure 3.5). As shown in Figure ES.1, projected changes in pollutant levels, primarily over Asia, may significantly increase surface temperature and reduce rainfall over the summertime continental United States throughout the second half of the twenty-first century (see Section 1.4 for details of the calculations).

2. The geographic patterns of factors that drive climate change due to short-lived gases and particles and the patterns of the resulting surface temperature responses are quite different. This is clearly seen in Figure ES.2, in which the largest fractional contribution to summertime radiative forcing from changes in short-lived pollutants in the last part of the twenty first century is primarily located over Asia, while the strongest warming response is located over the central United States. Regional emissions control strategies for short-lived pollutants will thus have global impacts on climate. The geographic disconnect between this driver of climate change and the surface temperature response is already apparent by 2050, as discussed in Section 3.3.4 and demonstrated in Figure 3.8.

3. Reductions of short-lived pollutants from the domestic fuel burning sector in Asia, whose climate impacts in this study (Section 3.4) are dominated by black carbon (soot), appear to offer the greatest potential for substantial, simultaneous

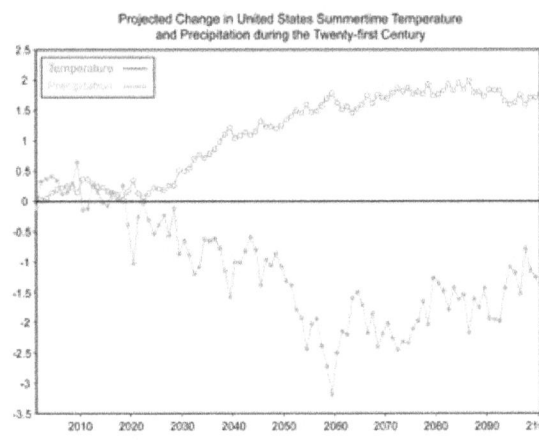

Figure ES.1 Calculation of twenty-first century temperature and precipitation change over the United States in summer (June through August) due to changes in short-lived gases and particles (see Section 1.4 for details). Temperature change in red is shown as the difference, in degrees Centigrade, from the 2001 value. Precipitation change in green is shown as the difference, in centimeters, from the 2001 value, and represents the sum of daily changes over the three months. The plotted data are 11 year moving averages.

improvement in local air quality and reduction of global warming. Reduction in
emissions from surface transportation in
North America would have a similar impact.

4. The three comprehensive climate models[3],
their associated chemical composition
models[4] and their differing projections of
short-lived emissions all lead to a wide range
of projected changes in climate due to short-
lived gases and particles. Each of the three
studies in this report represents a thoughtful,
but incomplete characterization of the driving forces and processes that are believed to
be important to the climate or to the global
distributions of the short-lived gases and
particles. Much work remains to be done to
characterize the sources of the differences
and their range. The two most important
uncertainties are found to be the projection
of future emissions and the determination of
the indirect effect[5] of particles on clouds. The
fundamental difference between uncertainties
in projecting future emissions and uncertainties in processes, such as the indirect effects
of particles, is discussed in Section 4.3.

5. The range of plausible short-lived emissions projections is very large, even for a
single well-defined global emission scenario (see Figures and discussion in Section
3.2 for details). Figure ES.3 clearly demonstrates this situation for the different
emission projections of black carbon particles (soot) used by the three research
groups. This currently limits our ability
to provide definitive statements on their
contribution to future climate change.

6. Natural particles such as dust and sea
salt also play an important role in climate
and their emissions and interactions differ significantly among the models, with
consequences to the role of short-lived

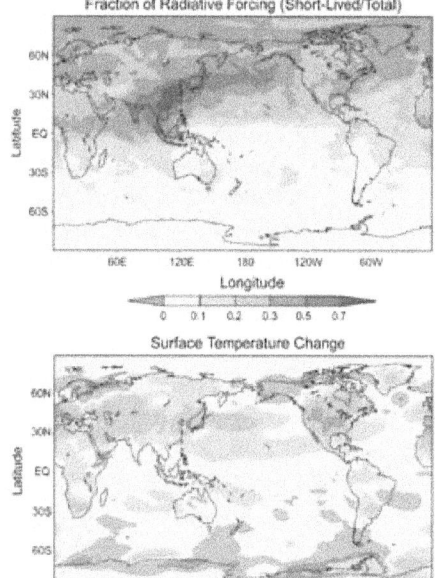

Figure ES.2 The fraction of summertime (June-August) radiative forcing* due to changing levels of short-lived gases and
particles and the resulting summertime surface temperature
change (degrees Centigrade) for year 2100.
*Radiative forcing is a measure of how the energy balance of the Earth-
atmosphere system is influenced when factors that affect climate, such
as atmospheric composition or surface reflectivity, are altered. When
radiative forcing is positive, the energy of the Earth-atmosphere system
will ultimately increase, leading to a warming of the system. In contrast,
for a negative radiative forcing, the energy will ultimately decrease, leading to a cooling of the system. For technical details, see Box 3.2.

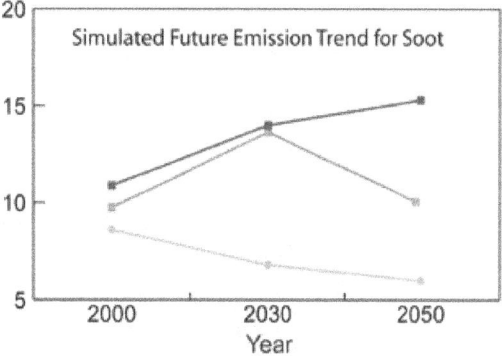

Figure ES.3 The three plausible but very different emis-
sions trends projected for black carbon particles (soot).
Each of the three groups in this study used a different trend.
The units are million metric tons of carbon per year.

---

[3] A comprehensive climate model is a state-of-the-art numerical representation of the climate based on the
physical, chemical, and biological properties of its components, their interactions, and feedback processes
that account for many of the climate's known properties. Coupled Atmosphere-Ocean (-sea ice) General
Circulation Models (AOGCMs) provide a comprehensive representation of the physical climate system.

[4] Chemical composition models are state-of-the-art numerical models that use the emission of gases and
particles as inputs and stimulate their chemical interactions, global transport by winds, and removal by rain,
snow, and deposition to the earth's surface. The resulting outputs are global three-dimensional distributions of the initial gases and particles and their products.

[5] Apart from the direct effects of particles absorbing and scattering radiation, particles produce an indirect
forcing of the climate system through their aiding in the formation of cloud droplets or by modifying the
optical properties and lifetime of clouds (see Box 3.1 for a more detailed discussion).

pollutants. This inconsistency among models should be addressed in future studies. This is discussed in Section 3.2.4 and demonstrated in Figure 3.2 and Table 3.5.

Regional emissions
control strategies
for short-lived
pollutants will have
large-scale impacts
on climate.

7. Climate projections based on new emissions scenarios where atmospheric concentrations of the long-lived greenhouse gases level off, or stabilize[6], at pre-determined levels (from CCSP SAP 2.1a) generally fall within the range of IPCC climate projections for the standard scenarios considered in the Fourth Assessment Report. The lower bound stabilization emissions scenarios, which have a carbon dioxide stabilization level of approximately 450 parts per million (by volume), result in global surface temperatures below those calculated for the lower bound IPCC scenario used in the Fourth Assessment, particularly beyond 2050. Nonetheless, all of them unequivocally cause warming across the range of possible emissions scenarios (see Section 2.6 for details).

[6] Stabilization emissions scenarios are a representation of the future emissions of a set of substances based on a coherent and internally consistent set of assumptions about the driving forces (such as population, socioeconomic development, and technological change) and their key relationships. These emissions are constrained so that the resulting atmospheric concentrations of the substance, or at least their net effect, level off at a predetermined value in the future.

ES.2 RECOMMENDATIONS FOR FUTURE RESEARCH

The five most critical areas for future research identified in this Report are:

1. The projection of future human-caused emissions for the short-lived gases and particles;
2. Indirect and direct effects of particles and mixing between particle types;
3. Transport, deposition, and chemistry of the short-lived gases and particles.
4. Regional climate forcing *vs.* regional climate response.
5. Sensitivity studies of climate responses to short-lived gases and particles.

1. Plausible emissions scenarios for the second half of the twenty-first century show significant climate impacts, yet the range of plausible scenarios is currently large and an increase in confidence in these scenarios is necessary. Short-lived gases and particles, unlike the well-mixed greenhouse gases, do not accumulate in the atmosphere. Therefore, combined with a large range of possible emissions scenarios, the climate impact of the short-lived gases and particles is currently extremely difficult to predict. Improvements in our ability to predict social, economic and technological developments affecting future emissions are needed. However, uncertainties in future emissions will always be with us. What we can do is develop a set of internally consistent emissions scenarios that include all of the important radiatively active gases and particles and bracket the full range of possible future outcomes.

2. The particle indirect effect (see Box 3.1 for a technical discussion), which is very poorly understood, is probably the process in most critical need of research. The climate modeling community as a whole cannot yet produce a credible characterization of the climate response to particle/cloud interactions. All models (including those participating in this study) are currently either ignoring it, or strongly constraining the model response. Attempts have been made using satellite and ground-based observations to improve the characterization of the

indirect effect, but major limitations remain and additional observations are required.

3. The three global composition models in this study all employed different treatments of mixing in the lowest layers of the atmosphere, transport and mixing by turbulence and clouds, removal of gases and particles by rain, snow and contact with the Earth's surface, and different approximate treatments of the very large collection of chemical reactions that we do not yet fully understand. Further coordinated model intercomparisons, evaluation of models against existing observations, and additional observations are all needed to achieve a better understanding of these processes.

4. The major unfinished analysis question in this study is the relative contribution of a model's regional climate response, as opposed to the contribution from the regional pattern of radiative forcing, to the simulated regional change in seasonal climate. Specific modeling studies are needed to answer questions such as: Is there a model-independent regional climate response? What are the actual physical mechanisms driving the regional surface temperature patterns that we observe? This appears to be a very important area of study, particularly given the strong climate response projected for the summertime central United States.

5. Analyses of surface temperature response to changes in short-lived gases and particles need to be strengthened by additional sensitivity studies that should help to clarify causes and mechanisms. There are also a wide range of

Partly as a result of the large range of possible future emissions scenarios, the climate impact of the short-lived gases and particles is currently extremely difficult to predict.

climate-chemistry feedbacks and controls that should be explored. Both the response of the climate system to controls on short-lived gases and particles and the possible feedbacks, and the possible impacts of climate changes on levels of short-lived gases and particles are all fertile areas for future research.

ES.3 GUIDE FOR READERS

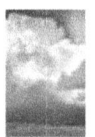

For those readers who would like to learn more about the research behind the Key Results and Findings and the Recommendations for Future Research, we provide the following guide to reading the four chapters. Chapter 1 provides an introduction to this study and relevant findings from previous climate research, introduces the goals and methodology, and provides Box 1.1 and Box 1.2 with relatively non-technical descriptions of the modeling tools and definitions of terms. It is written in a non-technical manner and is intended to provide all audiences with a general overview. Chapters 2 and 3 provide detailed technical information about specific models, model runs and projected trends and are intended primarily for the scientific community, though the key findings and the introduction to each chapter are written in non-technical language and intended for all audiences. Chapter 4 is intended for all audiences. It provides a summary of the major findings and identifies new opportunities for future research.

Introduction

Lead Authors: Hiram Levy II, NOAA/GFDL; Drew T. Shindell, NASA/GISS; Alice Gilliland, NOAA/ARL

Contributing Authors: Tom Wigley, NCAR; Ronald J. Stouffer, NOAA/GFDL

PROLOGUE

Comprehensive climate models[1] have become the essential tool for understanding past climates and making projections of future climate resulting from both natural and human causes. Projections of future climate require estimates (*e.g.*, scenarios) of future emissions of long-lived[2] greenhouse gases and short-lived[3] radiatively active[4] gases and particles. A number of standard emissions scenarios[5] have been developed for the Intergovernmental Panel on Climate Change (IPCC) assessment process, and the future impacts of these have been discussed extensively in the Fourth Assessment Report (IPCC, 2007).

As part of the Climate Change Science Program (CCSP) process, scenarios of long-lived greenhouse gas emissions, with the added requirement that their resulting atmospheric concentrations level off at specified values sometime after 2100 (*e.g.*, stabilization), were developed by the Synthesis and Assessment Product 2.1a team (Clarke *et al.*, 2007) and served as the basis for Synthesis and Assessment Product 3.2, for which the National Oceanic and Atmospheric Administration (NOAA) is the lead agency. NOAA's stated purpose for Synthesis and Assessment Product 3.2 is to provide information to those who use climate model outputs to assess the potential effects of human activities on climate, air quality, and ecosystem behavior. This report comprises two components that first assess the climate projections resulting from Synthesis and Assessment Product 2.1a scenarios in the context of existing IPCC climate projections, and then isolate and assess the future impacts on climate resulting from future emissions of short-lived gases and particles.

This second component explores the impact of short-lived radiatively active gases and particles on future climate, a critical issue that has recently become an active area of research in the reviewed literature (*e.g.*,

---

[1] Comprehensive climate models are a numerical representation of the climate based on the physical properties of its components, their interactions, and feedback processes. Coupled Atmosphere-Ocean (-sea ice) General Circulation Models (AOGCMs) represent our current state-of-the-art. However, they contain assumptions that may constrain or affect the accuracy and uncertainty of model output or results.

[2] Long-lived gases of interest have atmospheric lifetimes that range from ten years for methane to more than 100 years for nitrous oxide and carbon dioxide. Due to their long atmospheric lifetimes, they are well-mixed and evenly distributed throughout the lower atmosphere. Global atmospheric lifetime is the mass of a gas or particle in the atmosphere divided by the mass that is removed from the atmosphere each year.

[3] Short-lived radiatively active gases and particles of interest in the lower atmosphere have lifetimes of about a day for nitrogen oxides, a day to a week for most particles, and a week to a month for ozone. Their concentrations are highly variable and concentrated in the lowest part of the atmosphere, primarily near their sources.

[4] Radiatively active gases and particles absorb, scatter, and re-emit energy, thus changing the temperature of the atmosphere. They are commonly called greenhouse gases and particles.

[5] Emissions scenarios represent future emissions based on a coherent and internally consistent set of assumptions about the driving forces (*e.g.*, population change, socioeconomic development, technological change) and their key relationships.

Hansen *et al.*, 2000; Brasseur and Roekner, 2005; Delworth *et al.*, 2005). The existing state-of-the-art models used in this study represent incomplete characterizations of the driving forces and processes that are believed to be important to the climate responses and global distributions of the short-lived gases and particles. Moreover, these incomplete treatments are not consistent across the models. However, despite these challenges, this Report shows that short-lived gases and particles have a significant impact on climate, potentially throughout the twenty-first century.

1.1 HISTORICAL OVERVIEW

Climate models used for projections of the future have evolved substantially during the past several decades.

The climate models and the representation of the agents driving climate change used for projections of the future have both evolved substantially during the past several decades. In 1967, Manabe and Wetherald published the first model-based projection of future climate change. Using a simple model representing the global atmosphere as a single column, they projected a 2°C global surface air temperature change for a doubling of the atmospheric concentration of carbon dioxide. Model development continued on a wide range of numerical models, especially in the increasing sophistication of the ocean model.

In 1979, Manabe and Stouffer developed a global model at NOAA's Geophysical Fluid Dynamics Laboratory (GFDL) useful for estimating the climate sensitivity. They called this model an atmosphere-mixed layer ocean model, which is sometimes called a slab model. A slab model consists of global atmospheric, land, and sea ice component models, coupled to a static 50-meter deep layer of seawater. By construction, this type of model assumes no changes in the oceanic heat transports as the climate changes. It is used to estimate only equilibrium climate changes. In 1984, Hansen *et al.* used the NASA Goddard Institute of Space Studies (GISS) model in the first climate studies in which ocean heat transports were included in the climate calculation, although these were prescribed (fixed).

The two models discussed above, as well as one developed in part at the National Center for Atmospheric Research (NCAR), all played an important part in the first Intergovernmental Panel on Climate Change (IPCC) Assessment Report[6] (IPCC, 1990).

In the late 1980s, Washington and Meehl (1989) at NCAR and Stouffer *et al.* (1989) at GFDL developed the first comprehensive climate models (Atmosphere-Ocean General Circulation Models [AOGCMs]) useful for investigating climate change over multi-decadal and longer time periods. These models consisted of global atmosphere, ocean, land surface, and sea ice components. Both groups used an idealized radiative forcing to drive their models. Stouffer *et al.* used a one percent per year increase in the carbon dioxide concentration (compounded), where its atmospheric concentration doubles in 70 years.

By the time of the IPCC Second Assessment Report (IPCC, 1996), all three United States modeling centers were running comprehensive

---

[6] It should be noted that the IPCC does not directly perform any research. Rather, its reports are intended to be reviews of current research. However, it must also be noted that the IPCC is, in fact, a powerful driver of research and setter of research agendas in climate science. Moreover, only the latest report (Fourth Assessment Report) strictly enforced the requirement that all results discussed in it be previously published in the reviewed literature.

climate models. In addition, representation of the climate forcing was improving. Mitchell *et al.* (1995) in the United Kingdom (U.K.) developed a scheme for crudely incorporating the impact of sulfate particles on climate. Similarly, actual concentrations of long-lived greenhouse gases were used for the past, allowing more realistic climate simulations of the historical time period (1860 to present day). Using emissions scenarios developed by the IPCC in 1992 (IPCC, 1992), the U.K. group also made future projections of climate change through the year 2100. Their results were very important in the Second Assessment Report of the IPCC.

By the time of the Third IPCC Assessment Report in 2001 (IPCC, 2001), 12 comprehensive climate models were used to project climate out to the year 2100. They used the emissions scenarios produced by the Special Report on Emission Scenarios (Nakićenović and Swart, 2000), with most groups using a high (A2) and low (B2) emissions scenario. Some of the models included components to predict atmospheric particle concentrations, but most of the 12 models used variants of the Mitchell *et al.* (1995) method to include their impact on climate. While particle changes were included in the historical simulations, most of the future projections did not include any changes in them or tropospheric ozone.

In the most recent IPCC report, the Fourth Assessment Report of Working Group I (IPCC, 2007), 24 comprehensive climate models participated. The component models continued to become more sophisticated and included more physical processes. The new components allowed the inclusion of more radiatively active agents such as dust, black carbon and organic carbon particles, and land use in the scenarios. Again, most models included all or nearly all these climate forcing agents in their historical simulations, but many did not do so for the future. Most groups used the three standard scenarios developed for the IPCC by the Special Report on Emission Scenarios (Nakićenović and Swart, 2000) (B1, A1B and A2)[7] to make

their future climate projections. These are the same three scenarios represented in Figures 2.1 through 2.4 in Chapter 2.

1.2 GOALS AND RATIONALE

As described in the Prospectus outlining the purpose of this report, Synthesis and Assessment Product 3.2 has two primary goals:

1. Produce climate projections for research and assessment based on the stabilization emission scenarios of long-lived greenhouse gas emissions developed by Synthesis and Assessment Product 2.1a.
2. Assess the sign, magnitude, and duration of future climate changes due to changing levels of short-lived gases and particles that are radiatively active and that may be subject to future mitigation actions to address air quality issues.

The eight key questions which address the above goals, and which were also listed in the Prospectus for this report, are:

Q1. Do SAP 2.1a emissions scenarios differ significantly from IPCC emissions scenarios?

Q2. If the SAP 2.1a emissions scenarios do fall within the envelope of emissions scenarios previously considered by the IPCC, can the existing IPCC climate simulations be used to estimate 50-to 100-year climate responses for the SAP 2.1a carbon dioxide (CO_2) emissions scenarios?

In the most recent
IPCC report, the
Fourth Assessment
Report of Working
Group I, used 24
comprehensive
climate models.

---

[7] B1: emissions increase very slowly for a few more decades, then level off and decline; A2: emissions continue to increase rapidly and steadily throughout the twenty-first century; A1B: emissions increase very rapidly until 2030, continue to increase until 2050, and then decline.

Q3. What would be the changes to the climate system under the scenarios being put forward by SAP 2.1a?

Q4. For the next 50 to 100 years, can the time-varying behavior of the climate projections using the emissions scenarios from SAP 2.1a be distinguished from one another or from the scenarios currently being studied by the IPCC?

Q5. What are the impacts of the radiatively active short-lived species (gases and particles) not being reported in SAP 2.1a?

Q6. How do the impacts of short-lived species (gases and particles) compare with those of the well-mixed greenhouse gases as a function of the time horizon examined?

Q7. How do the regional impacts of short-lived species (gases and particles) compare with those of long-lived gases in or near polluted areas?

Q8. What might be the climate impacts of mitigation actions taken to reduce the atmospheric levels of short-lived species (gases and particles) to address air quality issues?

The answers to these questions are discussed in detail in Chapters 2 and 3.

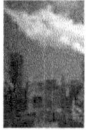

This Report focuses on the direct effect of short-lived pollutants on climate.

Synthesis and Assessment Product 3.2 is intended to provide information to those who use climate model outputs to assess the potential effects of human activities on climate, air quality, and ecosystem behavior. Since neither the IPCC nor SAP 2.1a explicitly addressed the direct influence of changing emissions of short-lived pollutants (carbon and sulfate particles and lower atmospheric ozone) on climate change, their role became a major focus of this report. This study encompasses a realistic time frame

over which available technological solutions can be employed, and focuses on those gases and particles whose future atmospheric levels are also subject to mitigation via air pollution control. Thus Synthesis and Assessment Product 3.2 can be very beneficial to all stakeholders of climate change science. The intended audiences include those engaged in scientific research, the media, policy makers, and members of the public. Policy and decision makers in the public sector (*e.g.*, congressional staff) need to understand the implications of these scenarios and the climates that they force, in contrast to the research science community, who may be more interested in the physical basis for the behavior.

1.3 LIMITATIONS

The first goal, assessing the climate projections for the SAP 2.1a stabilization emissions scenarios for long-lived greenhouse gases, is relatively narrowly defined and so treated. While the second goal, assessing the impact on climate of changing emissions of short-lived radiatively active gases and particles, could be viewed much more broadly, we do not. Our focus is primarily on the direct effect[8] of these short-lived pollutants on climate. Only in the case of methane do we explore any of the potential interactions of chemical sources, reactions, and removal resulting from a changing climate.

We do not examine any of the indirect effects[9] of pollutant particles on climate, nor do we address other potentially important impacts such as land use change, reactive nitrogen deposition and ecosystem responses, changing natural hydrocarbon emissions, changing oxidant levels and changing particle formation, or a wide range of other processes that can interact with climate, such as changes in ice clouds and vegetation burning. The resources were also not available for extensive sensitivity studies that might help explore more deeply the causes and mechanisms behind the potentially significant

---

[8] The direct effect refers to the influence of gases and particles on climate through scattering and absorbing radiation.

[9] Particles may lead to an indirect radiative forcing of the climate system by acting as cloud condensation nuclei or modifying the optical properties and lifetime of clouds.

of future climates resulting from natural and anthropogenic changes affecting the climate system. This whole modeling process is discussed in more detail in Box 1.1.

A number of standard scenarios have been developed for the IPCC assessment process, and the future impacts of these have been explored. As part of the Climate Change Science Program (CCSP) process, updated scenarios of long-lived greenhouse gases and their atmospheric concentrations were developed by the Synthesis and Assessment Product 2.1a team (Clarke *et al.*, 2007), and served as a basis for this Product. In addressing the first four questions, we examine the 12 scenarios for long-lived greenhouse gases developed by SAP 2.1a. We simulate the global surface temperature increases and sea-level rise (due only to thermal expansion of water, not melting ice caps) resulting from these scenarios using a simplified global climate computer model, MAGICC.

impact of short-lived pollutant levels on future climate. The above issues and many others are potential topics for future research, but were beyond the scope of this study. We will only address the impact on climate of direct radiative forcing by long- and short-lived greenhouse gases and particles.

1.4 METHODOLOGY

In addressing the questions posed above, we rely on several different types of computer models to project the climate changes that would result from the scenarios of emissions of greenhouse gases and particles. Projections of future climate first require estimates (*e.g.,* scenarios) of future emissions of long-lived greenhouse gases and radiatively active short-lived gases and particles. Next, global composition models (computer models of atmospheric transport and chemistry) employ the emissions scenarios to generate global distributions of the concentrations of short-lived radiatively active gases and particles. Then comprehensive climate models (computer models of the coupled atmosphere, land-surface, ocean, sea-ice system) employ global distributions of both the long-lived and short-lived radiatively active gases and particles to simulate past climates and make projections

In addressing the latter four questions listed in section 1.2, we focus on the effects of short-lived radiatively active gases and particles, and use three different state-of-the-art comprehensive climate models. Intercomparison studies, including the latest IPCC assessment, have shown that the performance of these models is comparable to other state-of-the-art comprehensive climate models (AOGCMs). Each of the three models was used to simulate future climate under two different scenarios: one in which human-caused short-lived gases and particles were allowed to change in the future, and one in which these gases and particles were held constant at present-day concentrations. The differences between the simulated climates for the two scenarios is attributed to the projected changes in the emissions of short-lived gases and particles.

1.5 TERMS AND DEFINITIONS

A number of technical terms are defined and briefly discussed for the benefit of those nontechnical readers who wish to proceed to Chapters 2 and 3. The definitions are collected in Box 1.2.

Each of the three models was used to simulate future climate under two different scenarios: one in which human-caused short-lived gases and particles were allowed to change in the future, and one in which these gases and particles were held constant at present-day concentrations.

BOX 1.1: Model Descriptions *(modified from IPCC Fourth Assessment Report)*

Integrated assessment models combine key elements of physical, chemical, biological, and economic systems into a decision-making framework with various levels of detail for the different components. These models differ in their use of monetary values, their integration of uncertainty, and their formulation of policy with regard to optimization, evaluation, and projections. For our study, their product was a set of stabilization emissions scenarios.

Chemical composition models are used to estimate the concentrations and distributions of trace gases and particles in the atmosphere that result from a given emissions scenario. These models, known technically as chemical transport models, are driven by winds, temperatures, and other meteorological properties that are either compiled from observations or supplied by climate models. Once the gas and particle emissions from human-induced and natural sources are supplied to the chemical composition model, they can be transported through the atmosphere, converted by chemical reactions, and removed from the atmosphere by rain, snow, and contact with the surface. These models provide concentrations of radiatively active gases and particles that vary in space and time for use in climate models.

A **climate model** is a numerical representation of the climate system based on the physical, chemical, and biological properties of its components, their interactions, and their feedback processes. The climate system can be represented by models of varying complexity. For any one component or combination of components, a hierarchy of models can be identified that differ in the number of spatial dimensions represented, the extent to which physical, chemical or biological processes are explicitly represented, or the level at which empirical parameterizations are involved.

Simple climate models estimate the change in global mean temperature and sea-level rise due to thermal expansion. They represent the ocean-atmosphere system as a set of global or hemispheric boxes, and predict global surface temperature using an energy balance equation, a prescribed value of climate sensitivity, and a basic representation of ocean heat uptake. Such models can also be coupled with simplified models of biogeochemical cycles, and allow rapid estimation of the climate response to a wide range of emissions scenarios. MAGICC (for details, see Appendix B) is such a coupled model.

State-of-the-art **comprehensive climate models** (generally referred to as AOGCMs, for Atmospere-Ocean General Circulation Models) include interacting components describing atmospheric, oceanic, and land surface processes, as well as sea ice. Although the large-scale dynamics of these models are treated exactly, approximations are still used to represent smaller, but still critical, processes such as the formation of clouds and precipitation, ocean mixing due to waves, and the mixing of air, heat, and moisture near the earth's surface. Uncertainties in these approximations are the primary reason for climate projections differing among different comprehensive climate models. Furthermore, the global models are generally unable to capture the small-scale features of climate in many regions. In such cases, the output from the global models can be used to drive regional climate models that have the same comprehensive treatment of interacting components, but, as they are only applied to part of the globe, are able to represent a region's climate in much greater detail.

"Storyline emissions scenario," and "stabilization emissions scenario," are two different approaches to estimating future emissions. The standard emissions scenarios used to provide the climate projections for the last two IPCC Assessment Reports (Third and Fourth) were storyline emissions scenarios. A set of economic development paths and rates of technological innovation, population growth, and social-political development were specified and Integrated Assessment Models (Box 1.1) were asked to solve for the greenhouse gas and particle emissions that were consistent with the specified conditions.

BOX 1.2: Useful Definitions

Storyline emissions scenarios are plausible representations of future emissions based on a coherent and internally consistent set of assumptions about the driving forces (e.g., population change, socioeconomic development, technological change) and their key relationships.

Stabilization emissions scenarios represent future emissions based on a coherent and internally consistent set of assumptions where, additionally, these emissions are constrained so that the resulting atmospheric concentration levels off at a predetermined value in the future.

Radiative forcing is a measure of how the energy balance of the Earth-atmosphere system is influenced when factors that affect climate are altered. The word radiative arises because these factors change the balance between incoming solar radiation and outgoing infrared radiation within the Earth's atmosphere. This radiative balance controls the Earth's surface temperature. The term "forcing" is used to indicate that Earth's radiative balance is being pushed away from its normal state. When radiative forcing from a factor or group of factors is evaluated as positive, the energy of the Earth-atmosphere system will ultimately increase, leading to a warming of the system. In contrast, for a negative radiative forcing, the energy will ultimately decrease, leading to a cooling of the system.

Global atmospheric lifetime is the mass of a gas or particle in the atmosphere divided by the mass that is removed from the atmosphere each year.

Long-lived gases of interest have atmospheric lifetimes that range from ten years for methane to more than 100 years for nitrous oxide. While carbon dioxide's lifetime is more complex, we can think of it as being more than 100 years in the climate system. As a result of their long atmospheric lifetimes, long-lived gases are well mixed and evenly distributed throughout the lower atmosphere. Their concentrations also change slowly with time.

Short-lived gases and particles of interest to climate studies have lifetimes of about a day for nitrogen oxides, a day to a week for most particles, and a week to a month for ozone. As a result of their short lifetimes, their concentrations are highly variable in space and time and are concentrated in the lowest part of the atmosphere, primarily near their sources.

Synthesis and Assessment Product 2.1a (Clarke *et al.*, 2007) took quite a different approach. They effectively established a set of targets for long-lived greenhouse gas concentrations, and then had their three integrated assessment models determine emissions pathways to those targets by applying economic principles to the relationships existing among economic development paths and rates of technological innovation, population growth, and social-political development. Each group used somewhat different approaches to determine the economic pathway to stabilization. Technically, only one of the models used the "least cost" approach in its strictest economic sense. However, as we show in Chapter 2, the resulting emissions and concentrations of the long-lived greenhouse gases over the twenty-first century are similar among models for a given target. Furthermore, all of the stabilization scenarios, with the exception of those for the lower bound target (only 18 percent increase in carbon dioxide over the next 100 years), fall within the range of the principal storyline scenarios used for the last two IPCC assessments. While the two approaches to constructing the emissions scenarios are different, the resulting concentrations of greenhouse gases, and their impacts on climate, are quite similar.

Radiative forcing is an important quantity that is frequently used when discussing the impact of radiatively active gases and particles. A technical definition is provided in Chapter 3, Box 3.2. We provide a relatively non-technical explanation in Box 1.2 of this chapter. It will be useful in the following discussion of long- and short-lived gases and particles.

The long-lived greenhouse gases have atmospheric lifetimes ranging from a decade to more than a century. As a result, they are uniformly mixed and their radiative forcing is also relatively uniformly distributed, both in space and time, throughout the lower atmosphere. On the other hand, the short-lived gases and particles have atmospheric lifetimes ranging from a day to weeks. Their concentrations are highly variable in space and time, and they are concentrated in the lowest part of the atmosphere, primarily near their sources. As a result their radiative forcing is also highly localized and can vary significantly in time. However, one of our Key Findings is that, while radiative forcing patterns for long- and short-term gases and particles are quite different, the regional patterns of climate change due to long- and short-lived radiatively active gases and particles are similar.

For those wishing to read further, we provide a brief reader's guide. Chapters 2 and 3 provide detailed technical information about specific models, model runs, and trends, and are intended primarily for the scientific community, though the Questions and Answers and the Introduction sections of each chapter are written in nontechnical language and are intended for all audiences. Chapter 4 is intended for all audiences. It provides a summary of the major findings, and identifies new opportunities for future research.

The regional patterns of climate change due to long- and short-lived gases and particles are similar, despite their different concentrations and lifetimes.

CHAPTER 2

Climate Projections from Well-Mixed Greenhouse Gas Stabilization Emission Scenarios

Lead Authors: Hiram Levy II, NOAA/GFDL; Drew T. Shindell, NASA/GISS; Alice Gilliland, NOAA/ARL

Contributing Author: Tom Wigley, NCAR

QUESTIONS AND ANSWERS

This chapter focuses on climate projections for the long-lived greenhouse gas stabilization emissions scenarios for the time period 2000 to 2100 that were produced under the U.S. Climate Change Science Program by an earlier Synthesis and Assessment Product, 2.1a (Clarke et al., 2007). Those scenarios[1] are called "stabilization emissions scenarios" because they are constrained so that the atmospheric concentrations of the long-lived greenhouse gases level off, or stabilize, at predetermined levels by the end of the twenty-first century. Our overall goal in this Chapter is to assess these "stabilization emissions scenarios" and the climates they would project for the twenty-first century in the context of the most recent Intergovernmental Panel on Climate Change Report, the Fourth Assessment Report of Working Group I (IPCC, 2007a). The major conclusions are summarized below as the answers to the first four questions in our Prospectus, and then receive more detailed attention in the remainder of the Chapter:

Q1. Do the stabilization emissions scenarios produced by Synthesis and Assessment Product (SAP) 2.1a differ significantly from those used in the Fourth Assessment Report of the IPCC?

A1. While different in concept and method of derivation (stabilization vs. "storyline"—see Box 1.2 for details) the long-lived greenhouse gas stabilization emissions scenarios outlined in Synthesis and Assessment Product 2.1a fall among the principal storyline emissions scenarios studied in the Fourth Assessment Report of the IPCC. While each individual stabilization emissions scenario differs somewhat from the individual IPCC scenarios, they are generally encompassed by the IPCC envelope of estimated future emissions.

Q2. If the Synthesis and Assessment Product 2.1a emissions scenarios do fall within the envelope of emissions scenarios previously considered by the IPCC, can the existing IPCC climate simulations be used to estimate 50 to 100 year climate responses for the SAP 2.1a carbon dioxide emissions scenarios?

A2. Given the close agreement between the ranges of emissions scenarios, time evolution of global concentrations and associated radiative forcings[2], and global mean temperature responses in the two assessments, we conclude that the key global and regional climate features noted in the IPCC reports can indeed be used to estimate the 50 to 100 year climate responses for the SAP 2.1a scenarios.

---

[1] Scenarios are representations of the future development of emissions of a substance based on a coherent and internally consistent set of assumptions about the driving forces (such as population, socioeconomic development, technological change) and their key relationships.

[2] Radiative forcing is a measure of how the energy balance of the Earth-atmosphere system is influenced when factors that affect climate, such as atmospheric composition or surface reflectivity, are altered. When radiative forcing is positive, the energy of the Earth-atmosphere system will ultimately increase, leading to a warming of the system. In contrast, for a negative radiative forcing, the energy will ultimately decrease, leading to a cooling of the system. For technical details, see Box 3.2.

Q3. What would be the changes to the climate system under the scenarios being put forward by SAP 2.1a?

A3. The key climate changes resulting from the "stabilization emissions scenarios" should be quite similar to the key findings from Chapters 10 (Meehl *et al.*, 2007) and 11 (Christensen *et al.*, 2007) of the Fourth Assessment Report of the IPCC, which are listed in Box 2.1 in Section 2.7 and discussed in more detail in Appendix A. The simulations by the simple climate model used in this Chapter, as well as the comprehensive climate model[3] simulations in Chapter 10 of the Fourth Assessment Report of the IPCC all find increases in global-average surface air temperature throughout the twenty-first century; with the warming increasing roughly proportional to the increasing concentrations of long-lived greenhouse gases.

Q4. For the next 50 to 100 years, can the climate projections using the emissions scenarios from SAP 2.1a be distinguished from one another or from the scenarios recently studied by the IPCC?

A4. For the first 30 years there is little difference in the predicted global-average climate among either the principal IPCC scenarios or the SAP 2.1a stabilization emissions scenarios for the long-lived greenhouse gases. For the second half of the twenty-first century, global mean and certain robust regional properties predicted for the different IPCC emission scenarios and applicable to the SAP 2.1a scenarios are distinguishable from each other in magnitude (the greater the concentration of long-lived greenhouse gases, the greater the magnitude) though not in their qualitative features.

---

[3] A comprehensive climate model is a numerical representation of the climate based on the physical, chemical, and biological properties of its components and their interactions and feedback processes, which account for many of its known properties. Coupled Atmosphere-Ocean (-sea ice) General Circulation Models (AOGCMs) provide the current state-of-the-art representation of the physical climate system.

2.1 INTRODUCTION

Chapter 2 is focused on climate projections for the four long-lived greenhouse gas scenarios developed by an earlier report, Synthesis and Assessment Product 2.1a (SAP 2.1a) (Clarke *et al.*, 2007). Our work in this chapter involves two different types of models:

1. Three integrated assessment models[4] that were used in Synthesis and Assessment Product 2.1a to produce stabilization emissions scenarios for long-lived greenhouse gases;
2. A simplified global climate model, Model for the Assessment of Greenhouse-gas Induced Climate Change (MAGICC)[5] that was used to simulate global levels of carbon dioxide, global-average radiative forcings for a variety of radiatively active[6] gases and particles, global-average surface temperature increases and global-average sea-level rise (due only to thermal expansion of water, not melting ice caps) for the four stabilization emissions scenarios.

The second section, 2.2, introduces the stabilization emissions scenarios and the models that were used to generate them in Synthesis and Assessment Product 2.1a. The stabilization levels were defined in terms of the combined radiative forcing for carbon dioxide (CO_2) and the other long-lived greenhouse gases that are potentially controlled under the Kyoto Protocol (methane, nitrous oxide, a suite of halocarbons, and sulfur hexafluoride [SF_6]). These radiative forcing levels were chosen to be more or less equivalent to 450, 550, 650, and 750 parts per

million (ppm) of carbon dioxide, and attainment was required within 100 to 200 years. For reference, preindustrial levels were approximately 280 ppm, and current levels of carbon dioxide are around 380 ppm.

Each integrated assessment model produced its own reference scenario, which is considered a "business as usual" or no-climate-policy scenario, as well as four stabilization emissions scenarios for long-lived greenhouse gas emissions that required a range of policy choices. The scenarios generated by each integrated assessment model were internally consistent, and each modeling group made independent choices in determining both their reference emissions, and their multi-gas policies required to achieve the specified stabilization levels. "All of the groups developed pathways to stabilization targets designed around economic principles. However, each group used somewhat different approaches to stabilization emissions scenario construction".

The third section, 2.3, introduces the simplified global climate model, MAGICC, which is used to generate the projections of carbon dioxide concentrations, radiative forcings due to the long-lived greenhouse gases, and global surface temperature increases for the four stabilization emissions scenarios introduced in the previous section 2.2. While the three integrated assessment models used in Synthesis and Assessment Product 2.1a each treated the cycling of carbon dioxide between the land, ocean and atmosphere in their own ways, in this study we use the carbon cycling treatment employed by MAGICC for all of the stabilization emissions scenarios. This provides a level playing field for all of the scenarios (see Wigley *et al.*, 2008 for a detailed discussion of this issue). We find that there is little difference between the two approaches.

MAGICC has four atmosphere boxes, one each over land and sea in each hemisphere, and two ocean boxes, one for each hemisphere. It consists of two highly simplified components: a climate component that has been adjusted to produce a global-average temperature change when the carbon dioxide concentration is doubled that is similar to the comprehensive climate models used in the IPCC Fourth As-

> Preindustrial levels of carbon dioxide were approximately 280 parts per million, and current levels are around 380 parts per million. The stabilization emissions scenarios used here were constructed to be more or less equivalent to 450, 550, 650, and 750 parts per million of carbon dioxide.

---

[4] Integrated assessment models are a framework of models, currently quite simplified, from the physical, biological, economic, and social sciences that interact among themselves in a consistent manner and can evaluate the status and the consequences of environmental change and the policy responses to it.

[5] MAGICC is a two-component numerical model consisting of a highly simplified representation of a climate model coupled with an equally simplified representation of the atmospheric composition of radiatively active gases and particles. This model is adjusted, based on the results of more complex climate models, to make representative predictions of global mean surface temperature and sea-level rise.

[6] "Radiatively active" indicates the ability of a substance to either absorb or emit sunlight or infrared radiation, thus changing the temperature of the atmosphere.

integrated assessment models did produce emissions scenarios for the short-lived pollutants that were consistent with the energy and policy decisions required for stabilization of the long-lived greenhouse gas concentrations. To assign a full radiative forcing to the scenarios calculated for the third model, an intermediate IPCC emissions scenario for the short-lived pollutants was added to its stabilization emissions scenario for long-lived gases. Again we find that the total radiative forcing (short-lived and long-lived radiatively active gases and particles) calculated by MAGICC for the 12 stabilization emissions scenarios fall among the total radiative forcings calculated by MAGICC for the principal storyline emissions scenarios employed in the Fourth Assessment Report of the IPCC.

Projected warming in the twenty-first century shows scenario-independent geographical patterns similar to those observed over the past several decades.

sessment Report, and a greenhouse gas and particle component that has also been adjusted to reproduce the global-average surface temperature and sea-level rise simulated by the same set of complex climate models for the various storyline emissions scenarios analyzed in the Fourth Assessment Report of the IPCC. A more detailed description of MAGICC is provided for the technical audience in Appendix B.

The fourth section, 2.4, shows that the concentrations of carbon dioxide projected by MAGICC for the twelve stabilization emissions scenarios (three models, four stabilization levels each) from Synthesis and Assessment Product 2.1a fall among earlier projections of carbon dioxide concentrations for the three primary storyline emissions scenarios employed in the Fourth Assessment Report of the IPCC (IPCC, 2007a). Next, it is shown that the 12 time series of radiative forcings for the long-lived greenhouse gases potentially regulated by the Kyoto Protocol, again calculated by MAGICC, fall among the time series of radiative forcings for the twenty-first century previously calculated for the same gases with the three time series of principal storyline emissions scenarios used in the IPCC Fourth Assessment Report.

The fifth section, 2.5, deals with the contribution of the short-lived pollutants (ozone, elemental and organic carbon particles and sulfate particles) to radiative forcing calculations by MAGICC for the stabilization emissions scenarios. While short-lived pollutants were not explicitly included in determining the stabilization emissions scenarios for the long-lived greenhouse gases, two of the three

The sixth section, 2.6, compares two sets of global-average surface temperature time series: an average of those calculated by a broad collection of comprehensive global climate models for the three principal IPCC emissions scenarios and reported in Chapter 10 of the IPCC's Fourth Assessment Report (Meehl *et al.*, 2007), and those calculated by MAGICC for the 12 SAP 2.1a stabilization emissions scenarios and reported here. As was found for the carbon dioxide concentration and radiative forcing time series discussed previously, the global-average surface temperatures calculated for the 12 stabilization emissions scenarios by MAGICC are generally contained within those calculated for the three IPCC scenarios by comprehensive global climate models. The exceptions are for the lower bound stabilization emissions scenario that would require carbon dioxide not to exceed 450 ppm by year 2100 (remember that current levels of carbon dioxide already exceed 380 ppm). The global-average surface temperatures tend to fall below those for the lowest IPCC scenario, particularly in the second half of the twenty-first century.

The seventh and final section, 2.7, addresses the primary objective of Chapter 2, *"Climate Projections for SAP 2.1a Scenarios."* While the stabilization emissions scenarios used in this report were derived in a fundamentally different manner from the storyline emissions scenarios used in the IPCC Fourth Assessment Report, they are generally contained within the storyline emissions scenarios and show a similar evolution with time. Moreover, the same is true for the resulting radiative forcings and global-average surface temperatures that are calculated with a simple global climate model. Drawing on the conclusion from the latest IPCC Summary for Policy Makers (IPCC, 2007b) that "Projected warming in the twenty-first century shows scenario-independent geographical patterns similar to those observed over the past several decades," we conclude that the robust conclusions arrived at in the latest IPCC report apply equally well to the climate responses expected for the four stabilization emissions scenarios provided by Synthesis and Assessment Product 2.1a.

2.2 WELL-MIXED GREENHOUSE GAS EMISSIONS SCENARIOS FROM SAP 2.1A

The three integrated assessment models used in SAP 2.1a were EPPA (Paltsev *et al.*, 2005), MiniCAM (Kim *et al.*, 2006) and MERGE (Richels *et al.*, 2007). These models have different levels of complexity in their modeling of socioeconomic, energy, industry, transport, and land-use systems. With respect to emissions, EPPA and MiniCAM are similarly comprehensive, and produce output for emissions of the following: all the major greenhouse gases (carbon dioxide [CO_2], methane [CH_4], nitrous oxide [N_2O], and a suite of halocarbons and sulfur hexafluoride [SF_6]); sulfur dioxide (SO_2), black carbon (BC) and organic carbon (OC) particles and their precursors; and the reactive gases carbon monoxide (CO), nitrogen oxides (NOx) and volatile organic compounds (VOCs), which are important determinants of tropospheric ozone change. MERGE produces emissions output for the major greenhouse gases and idealized short-lived and long-lived halocarbons (characterized by HFC-134a and SF_6), but not for any other short-lived radiatively active gases and particles and their precursors.

The stabilization levels were defined in terms of the combined radiative forcing for CO_2 and for the other gases that are potentially controlled under the Kyoto Protocol (CH_4, N_2O, halocarbons, and SF_6). *All of the groups developed pathways to stabilization targets designed around economic principles. However, each*

group used somewhat different approaches to stabilization emission scenario construction. (Reilly *et al.*, 1999; Manne and Richels, 2001; Sarofim *et al.*, 2005).

Consistent time series for the emissions of short-lived radiatively active gases and particles, carbon (both elemental and organic) and the precursors of sulfate particles and tropospheric ozone, were produced by the integrated assessment models to varying degrees, but the resulting radiative forcings were not part of the scenario definitions, nor were they considered as contributing to the radiative forcing targets. The stabilization levels for radiative forcing were constructed by determining the CO_2-only forcing associated with concentrations of 450, 550, 650, and 750 ppm and then adding additional radiative forcing to account for the other Kyoto Gases (0.8, 1.0, 1.2 and 1.4 W per m^2 respectively). The four stabilization levels are referred to as Level 1, Level 2, Level 3, and Level 4, where Level 1 requires the largest reduction in radiative forcing and is associated with CO_2 stabilization at roughly 450 ppm.

As SAP 2.1a (Clarke *et al.*, 2007) notes, "The three models display essentially the same relationship between greenhouse gas concentrations and radiative forcing, so the three reference scenarios also all exhibit higher radiative forcing, growing from roughly 2.2 W per m^2 above preindustrial in 2000 for the Kyoto Gases to between 6.4 W per m^2 and 8.6 W per m^2 in 2100". These differences arise primarily from differences in the assumptions underlying the reference scenarios, which lead to different reference emissions across the models.

The three models incorporate carbon cycles of different complexity, ranging from MERGE's neutral biosphere assumption to EPPA's coarse 3-D ocean. MiniCAM uses MAGICC to represent its carbon cycle. However, SAP 2.1a notes that the concentration of gases that reside in the atmosphere for long periods of time—decades to millennia—is more closely related to cumulative emissions than to annual emissions. In particular, this is true for CO_2, the gas responsible for the largest contribution to radiative forcing. This relationship can be seen for CO_2 in Figure 3.21 in SAP 2.1a (Clarke *et al.*, 2007), where cumulative emissions over the period

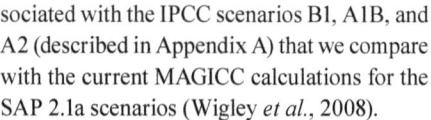

2000 to 2100, from the three reference scenarios and the 12 stabilization emission scenarios, are plotted against the CO_2 concentration in the year 2100. The plots for all three models lie on essentially the same line, indicating that despite considerable differences in representation of the processes that govern CO_2 uptake, the aggregate response to increased emissions is very similar. This basic linear relationship also holds for other long-lived gases, such as N_2O, SF_6, and the halocarbons."

The remainder of this chapter starts with the emissions scenarios generated by the three integrated assessment models in SAP 2.1a and examines their atmospheric composition, radiative forcing, and global-mean temperature. In the SAP 2.1a results, differences arise due to inter-model differences in the emissions for any given scenario, and differences between the models in their gas-cycle and climate components. Here we eliminate the second factor by using a single coupled gas-cycle/climate model to assess the scenarios—the MAGICC model as used in the IPCC Third Assessment Report (Cubasch and Meehl, 2001; Wigley and Raper, 2001). Many of the results given here have also been produced by the integrated assessment models, and some are described in SAP 2.1a. Using a single gas-cycle/climate model allows us to isolate differences arising from emissions scenario differences. Moreover, the MAGICC model was used previously to generate the carbon dioxide concentrations, Kyoto Gas[7] radiative forcing, and total radiative forcing as-

---
[7] "Kyoto Gases" refers to those long-lived greenhouse gases covered by the Kyoto Protocol (carbon dioxide, methane, nitrous oxide, hydrofluorocarbons, perfluorocarbons, and sulfur hexafluoride).

sociated with the IPCC scenarios B1, A1B, and A2 (described in Appendix A) that we compare with the current MAGICC calculations for the SAP 2.1a scenarios (Wigley *et al.*, 2008).

2.3 SIMPLIFIED GLOBAL CLIMATE MODEL (MAGICC)

MAGICC is a coupled gas-cycle/climate model that was used in the Third Assessment Report (Cubasch and Meehl, 2001; Wigley and Raper, 2001). A critical assessment focused on its skill in predicting global average sea-level rise is found in Chapter 10, Appendix 1 of the Working Group I contribution to the Fourth Assessment Report of the IPCC (Meehl *et al.* 2007).

The climate component is an energy-balance model with a one-dimensional, upwelling-diffusion ocean. For further details of models of this type, see Hoffert *et al.* (1980) and Harvey *et al.* (1997). In MAGICC, the globe is divided into land and ocean "boxes" in both hemispheres in order to account for different thermal inertias and climate sensitivities over land and ocean, and hemispheric and land/ocean differences in forcing for short-lived gases and particles such as tropospheric ozone and sulfate particles.

The climate model is coupled interactively with a series of gas-cycle models for CO_2, CH_4, N_2O, a suite of halocarbons, and SF_6. The carbon cycle model includes both CO_2 fertilization and temperature feedbacks, with model parameters tuned to give results consistent with the other carbon cycle models used in the Third Assessment Report (Kheshgi and Jain, 2003) and the Bern model (Joos *et al.*, 2001). For sulfate particles, both direct and indirect forcings are included using forcing/emissions relationships developed in Wigley (1989, 1991), with central estimates for 1990 forcing values.

The standard inputs to MAGICC are emissions of the various radiatively important gases and various climate model parameters. These parameters were tuned so that MAGICC was able to emulate results from a range of complex global climate models called Atmosphere-Ocean General Circulation Models (AOGCMs) in the Third Assessment Report (see Cubasch and Meehl, 2001). We use a value of 2.6°C equilibrium global-mean warming for a CO_2

doubling, the median of values for the above set of AOGCMs (see Appendix B for additional details).

2.4 LONG-LIVED GREENHOUSE GAS CONCENTRATIONS AND RADIATIVE FORCINGS

Figure 2.1 compares the concentrations of the primary greenhouse gas, CO_2, calculated by MAGICC for the 12 SAP 2.1a stabilization emission scenarios with earlier calculations of CO_2 concentrations for B1, A1B and A2, the principal storyline emission scenarios reported in Appendix II of the IPCC's Third Assessment Report (IPCC, 2001). For the first 20 years there is little difference among the 12 SAP 2.1a scenarios due to the long CO_2 lifetime, although the extreme Level 1 (L1) scenarios start to separate noticeably by 2030. By year 2100, CO_2 concentrations for the MiniCAM and EPPA Level 1 scenarios have converged on values close to 450 ppm. For MERGE, the 2100 value is lower. CO_2 concentrations for Levels 2 through 4 (L2 through L4) start to spread in the second half of the twenty-first century, but remain approximately bound between B1 and A1B all the way to 2100. EPPA now has the lowest CO_2 for Levels 2 through 4. The CO_2 levels for the lower bound Level 1 scenario, which requires immediate reductions in CO_2 emissions followed by ever increasing reductions (see SAP 2.1a for details), remain substantially below those for B1.

Next, Figure 2.2 considers where the radiative forcing due to increasing Kyoto greenhouse gases in the 12 SAP 2.1a stabilization emission scenarios, again calculated by MAGICC, are plotted with the Kyoto Gas radiative forcing values taken from Appendix II in the Third Assessment Report (IPCC, 2001) for the B1, A1B, and A2 storyline emission scenarios. The evolution of the 12 radiative forcing time series over the twenty-first century is very similar to that of CO_2, in Figure 2.1, which should not be surprising. However, there are some differences. The EPPA values undershoot the stabilization target for Levels 2 through 4 because they are on a trajectory where radiative forcing stabilizes some time after 2100, although emissions were calculated only to 2100 (Clarke et al., 2007). For the Level 2, 3 and 4 stabilization cases, it is not

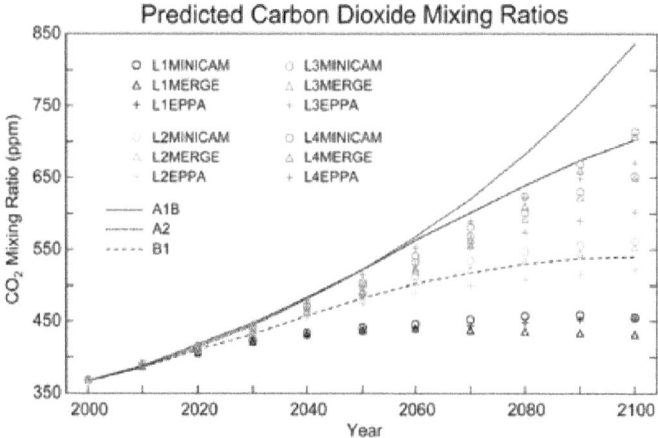

Figure 2.1 CO_2 concentrations (ppm) calculated by MAGICC for the 12 SAP 2.1a stabilization emission scenarios (Clarke et al., 2007) plotted with calculations of CO_2 concentrations for the principal scenarios (B1, A1B and A2) reported in Appendix II of the Third Assessment Report (IPCC, 2001b).

possible to stabilize as early as 2100 (Wigley et al., 1996). As we saw for carbon dioxide, the Kyoto Gas radiative forcing time series for stabilization Levels 2 through 4 are contained within the radiative forcings calculated for the IPCC scenarios, A1B and B1.

It should be noted that in general the three integrated assessment models hit their radiative forcing targets when they employ their own carbon cycle and atmosphere models. Thus, failure to hit these same radiative forcing targets when all three long-lived gases are run in MAGICC would seem to reflect the underlying uncertainties in the carbon cycles of the three integrated assessment models, which are known to be substantial.

For the first 20 years, there is little difference among the 12 Synthesized Assessment Product 2.1a scenarios, due to the long CO_2 lifetime.

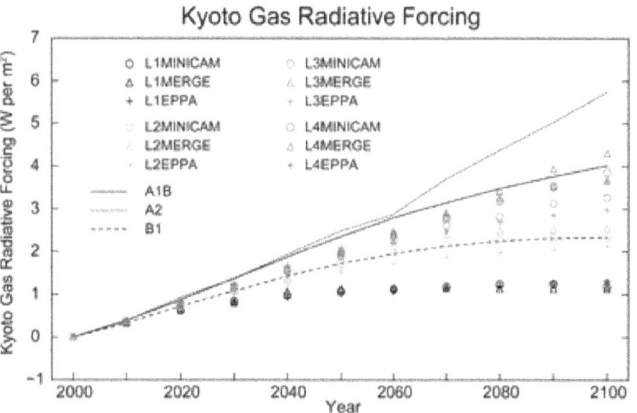

Figure 2.2 Kyoto Gas radiative forcing (W per m²) for the SAP 2.1a scenarios (Clarke et al., 2007), calculated by MAGICC, plotted with the Kyoto Gas Radiative Forcing values taken from Appendix II in the Third Assessment Report (IPCC, 2001) for the B1, A1B, and A2 SRES scenarios.

21

2.5 SHORT-LIVED GASES AND PARTICLES AND TOTAL RADIATIVE FORCING

The differences between the radiative forcing time evolution for the Kyoto Gases in Figure 2.2 and for all radiatively active gases and particles in Figure 2.3 are the result of differences among treatments of short-lived gases and particles.

While EPPA and MiniCAM produce emissions of sulfur dioxide (SO_2), elemental or black carbon and organic carbon particles and their precursors, and the key precursors of tropospheric ozone (CO, NOx and VOCs) as part of their model's climate projections, MERGE does not. To complete the MERGE scenarios, all four of its stabilization levels use the IPCC's B2 scenario of emissions for sulfur dioxide (Nakićenović and Swart, 2000) and assume that ozone precursor emissions remain constant. For all of the models, rather than use emissions for the elemental and organic particles, it is assumed that the elemental and organic particle radiative forcings track the sulfur dioxide emissions in each integrated assessment model's four stabilization emission scenarios. Therefore, while carbon dioxide emissions tend to track the IPCC scenarios, the emissions of short-lived gases and particles may be different, with the exception of sulfur dioxide emissions in MERGE.

Figure 2.3 compares the total radiative forcing calculated by MAGICC for the 12 SAP 2.1a scenarios, *i.e.*, the sum of Kyoto Gas forcings (Figure 2.2) plus forcings due to particles, tropospheric ozone, halocarbons controlled under the Montreal Protocol, and stratospheric ozone (Wigley *et al.*, 2008 and supplementary material referenced therein) with the total radiative

forcing calculated by MAGICC for the B1, A1B, and A2 scenarios used in the IPCC Fourth Assessment Report (IPCC, 2007a). Again, just as for CO_2 and Kyoto Gas radiative forcing, the 12 total radiative forcing time series do not begin to separate noticeably before 2030.

Because of the assumptions made about the short-lived gases and particles, the MERGE Kyoto Gas and total forcings differ least. MiniCAM shows the largest differences with total forcings now significantly exceeding the stabilization targets for all four levels, primarily due to sharp decreases in sulfur dioxide emissions, which produce significant increases in total radiative forcing by 2100 (approximately 1 W per m²). In the EPPA stabilization emission scenarios the changes in sulfur dioxide emissions are small, and most of the short-lived forcing comes from increased nitrogen oxide emissions that drive increases in tropospheric ozone and its positive radiative forcing (Wigley *et al.*, 2008). Remember that in SAP 2.1a, the stabilization targets were met using only the long-lived greenhouse gases.

The spread of stabilization forcings is significantly less for the Kyoto Gas forcings (which were used to define the stabilization targets) than for total forcing. Again the Level 1 total radiative forcings are generally below those of the B1 scenario, while the other Levels are bounded by B1 and A1B. However, in this case the Level 2 through 4 scenarios appear to track the B1 total radiative forcing out to 2060 to 2070 before the Level 3 and 4 scenarios start moving up to A1B. The differences between the radiative forcing time evolution for the Kyoto Gases in Figure 2.2 and for all radiatively active gases and particles in Figure 2.3 are the result of differences among treatments of short-lived gases and particles. The changes in global average surface temperatures that are driven by the total radiative forcing in Figure 2.3 are examined in the next section. We will continue to explore the potential impact of short-lived gases and particles on future global warming in considerable detail in Chapter 3.

Total Radiative Forcing

○ L1MINICAM	○ L3MINICAM
△ L1MERGE	△ L3MERGE
+ L1EPPA	+ L3EPPA
○ L2MINICAM	○ L4MINICAM
△ L2MERGE	△ L4MERGE
+ L2EPPA	+ L4EPPA
—— A1B	
···· A2	
- - - B1	

Figure 2.3 Total radiative forcing (W per m²) calculated by MAGICC for the 12 SAP 2.1a scenarios (Clarke *et al.*, 2007) plotted with the total calculated by MAGICC for the B1, A1B, and A2 scenarios (IPCC, 2001).

2.6 SURFACE TEMPERATURE: MAGICC AND IPCC COMPARISONS

Figure 2.4 compares multi-model global-mean surface temperature changes reported in Chapter 10 of the IPCC's Fourth Assessment Report for the standard storyline emission scenarios, B1, A1B and A2, with global-mean surface temperature changes calculated by MAGICC for the 12 SAP 2.1a stabilization emission scenarios (Clarke *et al.*, 2007). As we might expect, the general behavior is quite similar to that observed for total radiative forcing. All scenarios are close through 2020. Levels 2 through 4 stay in close agreement out to around 2050. The Level 1 scenarios are lower than B1, except for MiniCAM, where there is enhanced warming out to 2050 due to the rapid reduction in SO_2 emissions (Wigley, 1991). The other three levels follow B1 closely out to 2050 and then remain between B1 and A1B out to 2100.

For Level 1 and Level 2 temperatures, the rate of increase has begun to slow appreciably by 2100, which suggests that global-mean temperature could be stabilized if the emissions scenarios produced by the three integrated assessment models for these two stabilization cases (corresponding to 450 and 550 ppm CO_2, but also including the assumed or modeled levels of short-lived gases and particles) were followed. This in turn depends on the economic and technological feasibility of the Level 1 and 2 scenarios for both the long-lived greenhouse gases and the short-lived gases and particles. However, the temperatures for the less extreme Level 3 and 4 stabilization emission scenarios (corresponding to 650 and 750 ppm CO_2) are still growing, particularly Level 4 MiniCAM. It should also be noted that their upper bound, the A1B model-mean surface temperature, is also still growing at 2100. The global mean surface temperature projections for the 12 SAP 2.1a stabilization emission scenarios are well bounded by the comprehensive climate model simulations for the A1B scenario reported in Chapter 10 of the IPCC Fourth Assessment Report.

Table 2.1 displays the radiative forcings and temperature changes for the year 2100 for the 12 stabilization emissions scenarios from SAP 2.1a and for the three storyline emissions scenarios (A1B, A2 and B1) taken from the IPCC Fourth Assessment Report (IPCC, 2007a). These values were compiled from Figures 2.1 through 2.4.

2.7 CLIMATE PROJECTIONS FOR SAP 2.1A SCENARIOS

The 2.1a stabilization emissions scenarios (Clarke *et al.*, 2007) are derived in a fundamentally different manner from the development of the storyline emissions scenarios used in Fourth Assessment Report of the IPCC (IPCC, 2007a). However, we have shown in Section 2.4 that the 12 (three integrated assessment models, four stabilization emissions scenarios each) stabilization emissions scenarios reported in SAP 2.1a are contained within the range of the three principal storyline emissions scenarios used in the IPCC Assessment Report and show a similar evolution with time. The Kyoto Gases and total radiative forcings for those 12 emissions scenarios are generally constrained within the three principal scenarios used to make the climate projections discussed in Chapter 10 of the IPCC Fourth Assessment Report (Meehl *et al.*, 2007).

Section 2.6 shows that the global surface temperatures predicted for the SAP 2.1a scenarios over the twenty-first century by a simple coupled gas-cycle/climate model, MAGICC, fall within the range of the multi-model mean temperatures

Three integrated assessment models produced four stabilization emission scenarios each in Synthesized Assessment Product 2.1a. These 12 scenarios are generally contained within the range of the three principal storyline emission scenarios used in the IPCC Fourth Assessment Report and show a similar evolution with time.

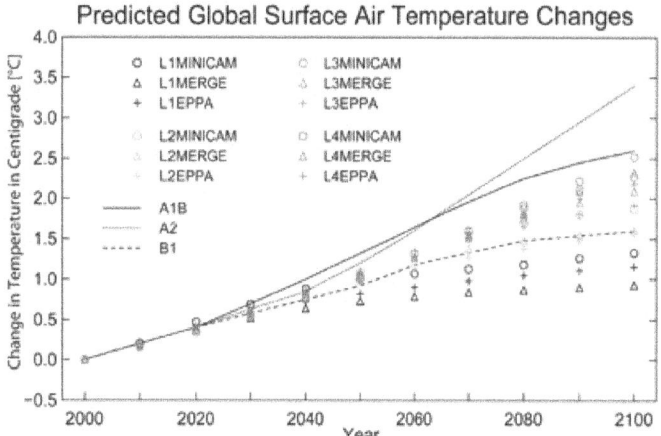

Figure 2.4 Multi-Atmosphere-Ocean General Circulation Model global-mean surface temperature changes (°C) reported in Chapter 10 of IPCC Fourth Assessment Report (Meehl *et al.*, 2007) for the standard storyline emissions scenarios B1, A1B, and A2 plotted with global-mean surface temperatures calculated by MAGICC for the 12 SAP 2.1a stabilization emissions scenarios (Clarke *et al.*, 2007).

We conclude that the robust conclusions arrived at in Chapter 10 of the Fourth Assessment Report regarding the predicted climate response to the three scenarios studied in the most detail in that report apply equally well to the climate responses expected for the four long-lived greenhouse gas stabilization emission scenarios provided by SAP 2.1a.

calculated with state-of-the-art comprehensive climate models for the three principal IPCC scenarios and reported in Chapter 10 (Meehl *et al.*, 2007). In fact, the global average surface temperatures for Levels 2 through 4 scenarios all track the values reported by the IPCC for B1 out to 2050. The primary exceptions are all of the Level 1 scenarios beyond year 2050 which are significantly below B1. We also draw on the conclusion in the Summary for Policy Makers in the IPCC Fourth Assessment Report (IPCC, 2007b): "Projected warming in the twenty-first century shows scenario-independent geographical patterns similar to those observed over the past several decades." Figure 10.8 in Chapter 10 of the Fourth Assessment Report (Meehl *et al.*, 2007) also clearly shows that the geographical pattern of the robust climate features are preserved across scenarios employed in the IPCC projections for the twenty-first century climate, while the magnitude of the warming increases with the magnitude of the radiative forcing and with increases in the concentration of the long-lived greenhouse gases.

We conclude that the robust conclusions arrived at in Chapter 10 of the Fourth Assessment Report (Meehl *et al.*, 2007) regarding the predicted climate response to the three scenarios studied in the most detail in that Report, (B1, A1, and A1B) apply equally well to the climate responses expected for the four long-lived greenhouse gas stabilization emission scenarios (three realizations of each) provided by SAP 2.1a (Clarke *et al.*, 2007). These robust conclusions are highlighted in Box 2.1 below and further discussed in Appendix A.

At this time, we also introduce in Box 2.2 our general approach to treating uncertainty in this document. Since much of this report deals with ranges of projections of radiative forcing and surface temperature rather than explicit predictions, we do not generally assign uncertainty values. We do quote the IPCC explicit uncertainty values in Box 2.1. Later in Chapter 3 we present Box 3.3 that addresses the determination of statistical significance and our use of it in a more technical manner.

Table 2.1 Year 2100 values from Figures 2.1, 2.2, 2.3, and 2.4.

Scenario	CO2 (ppm) (Figure 2.1)	Kyoto Gases Radiative Forcing (W per m²) (Figure 2.2)	Total (W per m²) (Figure 2.3)	Temperature Change (degrees C) (Figure 2.4)
A2	836	5.75	6.74	3.40
A1B	703	4.02	4.72	2.60
B1	540	2.34	2.86	1.60
L1 MiniCAM	454	1.17	2.04	1.32
L1 Merge	432	1.14	1.36	0.93
L1 EPPA	453	1.28	1.75	1.16
L2 MiniCAM	559	2.33	3.10	1.83
L2 Merge	553	2.56	2.71	1.61
L2 EPPA	551	2.12	2.58	1.56
L3 MiniCAM	651	3.23	4.09	2.27
L3 Merge	650	3.67	3.81	2.09
L3 EPPA	601	2.98	3.36	1.92
L4 MiniCAM	712	3.83	4.73	2.50
L4 Merge	708	4.30	4.45	2.33
L4 EPPA	668	3.63	3.97	2.18

BOX 2.1: Robust conclusions for global climate from Chapter 10 of the Fourth Assessment Report (Meehl et *al.*, 2007):

- Surface Air Temperatures show their greatest increases over land (roughly twice the global average temperature increase), over wintertime high northern latitudes, and over the summertime United States and southern Europe, and show less warming over the southern oceans and North Atlantic. These patterns are similar across the B1, A1B, and A2 scenarios with increasing magnitude with increasing radiative forcing.
- It is very likely that heat waves will be more intense, more frequent, and longer lasting in a future warmer climate.
- By 2100, global-mean sea level is projected across the 3 SRES scenarios to rise by 0.28m to 0.37m for the three multi-model averages with an overall 5-95 percent range of 0.19 to 0.50 m. Thermal expansion contributes 60-70 percent of the central estimate for all scenarios. There is, however, a large uncertainty in the contribution from ice sheet melt, which is poorly represented in current models.
- Globally averaged mean atmospheric water vapor content, evaporation rate, and precipitation rate are projected to increase. While, in general, wet areas get wetter and dry areas get dryer, the geographical patterns of precipitation change during the twenty-first century are not as consistent across the complex climate model simulations and across scenarios as they are for surface temperature.
- Multi-model projections based on SRES scenarios give reductions in ocean pH of between 0.14 and 0.35 units over the twenty-first century, adding to the present decrease of 0.1 units from preindustrial times.
- There is no consistent change in El Niño-Southern Oscillation (ENSO) for those complex climate models that are able to reproduce ENSO-like processes.
- Those models with a realistic Atlantic Meridional Overturning Circulation (MOC) predict that it is very likely that the MOC will slow by 2100, but will not shut down.
- The Fourth Assessment Report Summary for Policymakers finds it "Likely that intense hurricanes and typhoons will increase through the twenty-first century".

There are also important robust conclusions for North America from Chapter 11 of the Fourth Assessment Report (Christensen *et al.*, 2007):
- "All of North America is very likely to warm during this century, and the annual mean warming is likely to exceed the global-mean warming in most areas."
- "Annual-mean precipitation is very likely to increase in Canada and the U. S. Northeast, and likely to decrease in the U.S. Southwest".
- "Snow season length and snow depth are very likely to decrease in most of North America, except in the northernmost part of Canada where maximum snow depth is likely to increase".

NOTE: The terms "very likely" and "likely" have specific statistical meanings defined by the IPCC.

Very likely	greater than 90 percent chance of occurring
Likely	greater than 67 percent chance of occurring

BOX 2.2: Uncertainty

In doing any assessment, it is helpful to precisely convey the degree of certainty of various findings and projections. There are numerous choices for categories of likelihood and appropriate wording to define these categories. In Chapter 2 of this report, since many of the findings of this Report are comparable to those discussed in the Fourth Assessment Report of the IPCC, we have chosen to be consistent with the IPCC lexicon of uncertainty:

Lexicon	Probability of Occurrence
Virtually certain	> 99 percent
Extremely likely	> 95 percent
Very Likely	> 90 percent
Likely	> 66 percent
More likely than not	> 50 percent
Unlikely	< 33 percent
Very unlikely	< 10 percent
Extremely unlikely	< 5 percent

Elsewhere in the report, we are projecting climate, based on model simulations that use, as a foundation, scenarios of short-lived gases and particles, which are themselves plausible, but highly uncertain. For this reason, we have largely avoided assigning uncertainty values. However, where they do occur, we have condensed the IPCC ranges of uncertainty to fewer categories because we are unable to be as precise as in the IPCC assessments, which consider primarily the long-lived greenhouse gases. This lexicon is also consistent with other CCSP reports, such as SAP 3.3 and SAP 4.1.

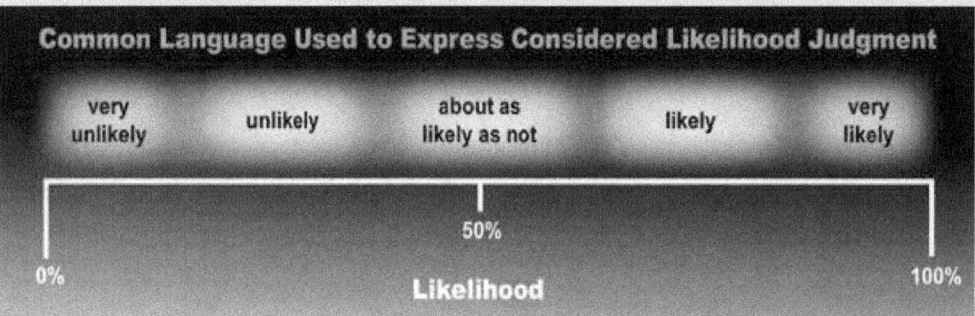

Figure Box 2.2 Language in this Synthesis and Assessment Product (Chapters 3 and 4) used to express the team's expert judgment of likelihood, when such a judgment is appropriate.

Climate Change from Short-Lived Emissions Due to Human Activities

Lead Authors: Drew T. Shindell, NASA/GISS; Hiram Levy II,
NOAA/GFDL; Alice Gilliland, NOAA/ARL; M. Daniel Schwarzkopf,
NOAA/GFDL; Larry W. Horowitz, NOAA/GFDL

Contributing Author: Jean-Francois Lamarque, NCAR

QUESTIONS AND ANSWERS

This chapter addresses the four questions regarding short-lived gases and particles that were posed in the Prospectus for this report:

Q5. What are the impacts of the radiatively active short-lived species (gases and particles) not explicitly the subject of prior CCSP assessments (SAP 2.1a: Scenarios of Greenhouse Gas Emissions and Atmospheric Concentrations)?

A5. Uncertainties in emissions projections for short-lived gases and particles are very large, even for a particular storyline. For particles, these uncertainties are usually dominant, while for tropospheric ozone, uncertainties in physical processes are more important. Differences among modeled future atmospheric burdens and radiative forcing for particles are dominated by divergent assumptions about emissions from South and East Asia. Particle mixing, particle indirect effects, the influence of ecosystem-chemistry interactions on methane, and stratosphere-troposphere exchange all contribute to large uncertainties separate from the emissions projections.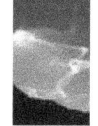

Q6. How do the impacts on climate of short-lived gases and particles compare with those of the well-mixed greenhouse gases as a function of the time horizon examined?

A6. By 2050 in two of the three models, changes in short-lived gas and particle concentrations (which are not reported in SAP 2.1a stabilization emission scenarios [Clarke et al., 2007]) contribute approximately 20 percent of global-mean annual-average warming, while one model shows virtually no effect. To a large extent, the inter-model differences are related to differences in emissions. Changes in the levels of short-lived gases and particles may play a substantial role out to 2100. One model finds that they can contribute 40 percent of the total projected summertime warming in the central United States over the second half of the 2100 century.

Q7. How do the regional impacts of short-lived species (gases and particles) compare with those of long-lived gases in or near polluted areas?

A7. The spatial distribution of radiative forcing is generally less important than the spatial distribution of climate response in predicting the impact on climate. Thus, both short-lived and long-lived gases and particles appear to cause enhanced climate responses in the same regions, rather than short-lived gases and particles having an enhanced effect primarily in or near polluted areas.

Q8. What might be the climate impacts of mitigation actions taken to reduce the atmospheric levels of short-lived species (gases and particles) to address air quality issues?

A8. Regional air quality emissions control strategies for short-lived pollutants have the potential to substantially affect climate globally. In one study, emissions reductions in the domestic energy/power sector in developing Asia, and to a lesser extent in the surface transportation sector in North America, appear to offer the greatest potential for substantial, simultaneous improvement in local air quality and mitigation of global warming.

3.1 INTRODUCTION

In this chapter, we describe results from numerical simulations of twenty-first century climate, with a major focus on the effects of short-lived gases and particles. The calculations incorporate results from three different types of models:

1. Integrated assessment models that produce emissions scenarios for particles and for ozone and particle precursors.
2. Global chemical composition models, which employ these emissions scenarios to generate concentrations for the short-lived radiatively active gases and particles.
3. Global comprehensive climate models, which calculate the climate response to the projected concentrations of both the short-lived and long-lived gases and particles. Box 1.1 outlines this sequence in detail.

The second part of Chapter 3, Section 3.2, is a discussion of the emissions scenarios and the models used to generate them, and the chemical composition models (sometimes called chemical transport models) used to produce the global distributions of short-lived gases and particles that help to drive the comprehensive climate models. Section 3.2 shows that, beginning with a single socioeconomic scenario for the time evolution of long-lived (well-mixed) greenhouse gases, different assumptions about the evolution of the particles and precursor gases lead to very different estimates of particle and ozone concentrations for the twenty-first century. We conclude that uncertainties in emissions projections for short-lived gases and particles are very large, even for a particular storyline. For particles, these uncertainties are usually dominant, while for tropospheric ozone, uncertainties in physical processes are more important.

The third part of Chapter 3, Section 3.3, discusses the three global comprehensive climate models (Geophysical Fluid Dynamics Laboratory [GFDL]; Goddard Institute for Space Studies [GISS]; Community Climate System Model [CCSM] developed in part at the National Center for Atmospheric Research [NCAR]) that have been used to calculate the impact of the

short- and long-lived gases and particles[1] on the climate, focusing on the changes in surface temperature and precipitation. Supplementing the climate model results are calculations of the changes in radiative forcing[2] of the earth-atmosphere system. We find that by 2050, two of the three climate models show that approximately 20 percent of the global-mean annual-average warming is due to changing levels of radiatively active short-lived gases and particles. One model shows virtually no effect from short-lived gases and particles. To a large extent, the inter-model differences are related to differences in emissions. An extensive discussion and comparison of the projected distributions of short-lived species and resulting radiative forcings employed by the three groups and the resulting climate projections from the three comprehensive climate models can be found in Shindell *et al.* (2008).

One of the models has been extended to 2100. In that model, short-lived gases and particles play a substantial role, relative to the well-mixed greenhouse gases, in the surface temperature evolution out to 2100 and are responsible for 40 percent of the projected 2100 summertime warming in the central United States (Levy *et al.*, 2008).

The fourth part of Chapter 3, Section 3.4, discusses the effects of changes in regional particle and ozone and particle precursor emissions, using models that separate emissions by economic sector. The results show that regional air quality emissions control strategies for short-lived pollutants have the potential to substantially affect climate at large-scales. Emissions reductions from domestic sources in Asia, and to a lesser extent from surface transportation in North

---

[1] We distinguish here between short-lived gases and particles (which have atmospheric lifetimes less than one month and are non-uniformly distributed) and long-lived gases (which have lifetimes of a decade or more and are generally well mixed in the atmosphere).

[2] Radiative forcing is a measure of how the energy balance of the Earth-atmosphere system is influenced when factors that affect climate, such as atmospheric composition or surface reflectivity, are altered. When radiative forcing is positive, the energy of the Earth-atmosphere system will ultimately increase, leading to a warming of the system. In contrast, for a negative radiative forcing, the energy will ultimately decrease, leading to a cooling of the system. For technical details, see Box 3.2.

America, appear to offer the greatest potential for substantial, simultaneous improvement in local air quality and mitigation of global climate change.

3.2 EMISSIONS SCENARIOS AND COMPOSITION MODEL DESCRIPTIONS

3.2.1 Emissions Scenarios

The long-lived (well-mixed) greenhouse gases included in this study were carbon dioxide (CO_2), nitrous oxide (N_2O), methane (CH_4), and the minor gases (chlorofluorocarbons, sulfur hexafluoride). Projected global mean values were prescribed following the A1B "marker" storyline emissions scenario for all three modeling groups. Emissions for anthropogenic sources of particles and precursor gases and particles for all three composition model calculations were based on an international emissions inventory maintained in the Netherlands (Olivier and Berdowski, 2001).

Though the three groups in this study all prescribed future emissions following a specific socioeconomic scenario (A1B) that was highly studied in the Fourth Assessment by the Intergovernmental Panel on Climate Change (IPCC, 2007), they used different emissions trends for the short-lived gases and particles. There are several reasons for the differences. For one, the A1B emissions projections only provide estimates of anthropogenic emissions, and each model used its own natural emissions (though these were largely held constant). Secondly, integrated assessment models, while using the same socio-economic storyline (A1B), provided a range of emissions results (Nakićenović and Swart, 2000).

Two groups, GFDL and NCAR, used output from the AIM integrated assessment model (integrated assessment models are defined in Chapter 2, Section 2.1) while GISS used results from the IMAGE model. Though the emissions output generated from AIM was denoted the "marker" scenario by the IPCC, it was noted that it did not represent the average, best, or median result, and that all integrated assessment model results should be treated equally. Finally, emissions for some gases and particles, such as carbonaceous particles, were not provided.

This last issue motivated the GISS choice of the IMAGE model output, as it provided sufficient regional detail to allow carbonaceous particle emissions to be estimated consistently with the other gases and particles. Another complexity was the treatment of biomass burning emissions, which are partly natural and partly anthropogenic. In the GFDL model, biomass-burning emissions were assumed to be half anthropogenic and half natural. The GISS model instead used biomass burning emissions projections from another inventory (Streets *et al.*, 2004).

The result is a substantial divergence in the projected trends among the three models (Figure 3.1, Table 3.1). For sulfur dioxide (SO_2), the precursor to sulfate particle, the emissions follow reasonably similar trajectories, with globally averaged increases until 2030 followed by decreases to 2050 and even further decreases to 2100. However, the percentage increase is roughly double for GISS and CCSM as compared with GFDL. Thus even two composition models using anthropogenic emissions projections from the same integrated assessment model show large differences in the evolution of their total emissions, presumably owing to differences in the present-day emissions inventories. At 2050, the GFDL model has substantially reduced emissions compared with 2000, while the other models show enhanced emissions relative to 2000. A similar divergence in projected sulfur-dioxide trends is present in the SAP 2.1a stabilization emissions scenarios (Clarke *et al.*, 2007) discussed in Chapter 2, with emissions decreasing dramatically (~70 percent) by

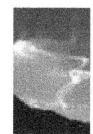

Though the three groups in this study all prescribed future emissions following a specific socioeconomic scenario, the scenario has multiple interpretations, and hence they used different emissions trends for the short-lived gases and particles.

Table 3.1 Global emissions. Emissions (in Teragrams, Tg = 1 x 10^{12} g) include both natural and anthropogenic sources. Values in parentheses are changes relative to 2000.

Species	Model	2000	2030	2050	2100
NO$_x$ (Tg N per year)	GFDL	40	57 (43%)	54 (35%)	48 (20%)
	GISS	50.5	67.0 (33%)	77.5 (53%)	NA
BC (Tg C per year)	GFDL	10.9	14.0 (28%)	15.3 (40%)	19.9 (83%)
	GISS	8.6	6.8 (-21%)	6.0 (-30%)	NA
OC (Tg C per year)	GFDL	51.5	61.9 (20%)	66.5 (29%)	84.3 (%)
	GISS	69.5	57.0 (-18%)	58.3 (-16%)	NA
SO$_2$ (Tg SO$_2$ per year)	GFDL	147	187 (27%)	118 (-20%)	56 (-62%)
	GISS	130	202 (55%)	164 (26%)	NA
	CCSM	125	190 (52%)	148 (18%)	NA
Dust (Tg per year)	GFDL	2471	2471	2471	2471
	GISS	1580	1580	1580	NA

2050 in one integrated assessment model (MINICAM) while decreasing only moderately (~20 percent) in the two others, and even beginning to increase again after about 2040 in one of those two.

Differences are even more striking for carbonaceous particle emissions, which were not provided by any of the integrated assessment models. We focus on black carbon (BC) as the more important radiative perturbation. For this particle (and for organic carbon [OC]), the GFDL composition model uses the IPCC recommendation to scale carbonaceous particle emissions to carbon monoxide emissions, leading to substantial increases with time (Figure 3.1, Table 3.1). However, many of the sources of carbon monoxide emissions are different from those of carbonaceous particles. The NCAR group did not simulate the future composition of black and organic carbon based on emissions projections, but instead scaled their present-day distribution by the global factors derived for sulfur dioxide. The time evolution of black and organic carbon emissions in the CCSM model thus follows the same trajectory as that of SO$_2$. On the other hand, the GISS group used emissions projections from Streets *et al.* (2004) based on energy and fuel usage trends from the IMAGE model (as for other gases and particles) and including expected changes in technology. This led to a substantial reduction in future emissions of carbonaceous particles.

For precursors of tropospheric ozone, there was again divergence among the models. The primary precursor in most regions, NO$_x$ (nitrogen oxides = NO + NO$_2$), increased steadily in the projections used by GISS, while it peaked at 2030 and decreased slightly thereafter in the projections used at GFDL (Table 3.1). Hydrocarbons and carbon monoxide show analogous differences. Methane was prescribed according to the A1B "marker" scenario values for all three composition models. Thus ozone, in addition

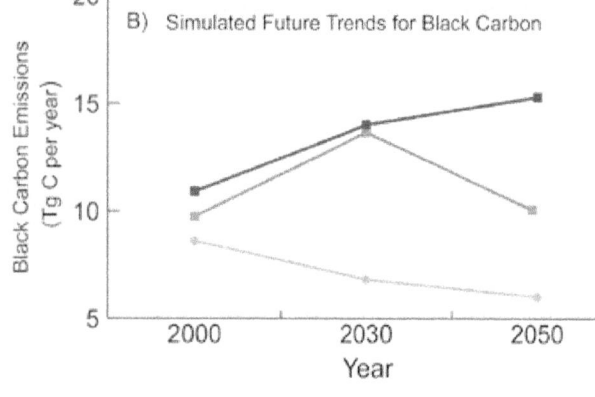

Figure 3.1 A1B emissions trends used in the three models for SO$_2$ (top) and black carbon (bottom). Note that in the CCSM model, the present day black carbon distribution was scaled in the future rather than calculated from black carbon emissions. Scaling was chosen to mimic the global sulfur dioxide emissions, a 40 percent increase over 2000 at 2030, and 10 percent at 2050. The CCSM 2000 black carbon global emission is set at the average of the GISS and GFDL 2000 values, and follows this scaling in the future, for illustrative purposes.

to the particles, was modeled in substantially different ways at the three centers.

The three models included projected changes in the same gases and particles, with the exception of nitrate, which only varied in the GISS model. As its contribution to total particle and total particle radiative forcing is small, at least in this GISS model, this particular difference was not significant in our results.

3.2.2 Composition Models

The chemical composition models used to produce short-lived gas and particle concentrations for the GFDL, GISS, and CCSM climate models were driven by the emissions projections discussed in Section 3.2.1. While the three models did not use identical present-day emissions, their anthropogenic emissions were based on the same international inventory (Olivier and Berdowski, 2001). The chemical composition simulations were run for one or two years, with the three-dimensional monthly mean concentrations and optical properties archived for use as off-line concentration fields to drive the climate model simulations discussed in Section 3.3. These simulations were all performed with present-day meteorology (values for temperature, moisture, and wind). Further details about the chemical composition models are provided in Appendix C.

3.2.2.1 GEOPHYSICAL FLUID DYNAMICS LABORATORY (GFDL)

Composition changes for the short-lived gases and particles in the GFDL experiments were calculated using the global chemical transport model MOZART-2 (Model for OZone And Related chemical Tracers, version 2.4), which has been described in detail previously (Horowitz *et al.*, 2003; Horowitz, 2006; and references therein). This model was used to generate the monthly average distributions of tropospheric ozone, sulfate, and black and organic carbon as a function of latitude, longitude, altitude, and time for the emissions scenarios discussed above. Simulated ozone concentrations agree well with present-day observations and recent trends (Horowitz, 2006). Overall, the predicted concentrations of particle are within a factor of two of the observed values and have a tendency to be overestimated (Ginoux *et al.*, 2006). Further details on the MOZART model are found

in Appendix C, in the section on Geophysical Fluid Dynamics Laboratory.

3.2.2.2 GODDARD INSTITUTE FOR SPACE STUDIES (GISS)

The configuration of the GISS composition model used here has been described in detail in (Shindell *et al.*, 2007). In brief, the composition model PUCCINI (Physical Understanding of Composition-Climate INteractions and Impacts) includes ozone and oxidant photochemistry in both the troposphere and stratosphere (Shindell *et al.*, 2006b), sulfate, carbonaceous and sea-salt particles (Koch *et al.*, 2006, 2007), nitrate particles (Bauer *et al.*, 2007), and mineral dust (Miller *et al.*, 2006a). Present-day composition results in the model are generally similar to those in the underlying chemistry and particle models. Further details on the PUCCINI model resolution, composition, and performance are found in Appendix C, in the section on Goddard Institute for Space Studies.

3.2.2.3 COMMUNITY CLIMATE SYSTEM MODEL (CCSM)

For the climate simulations described in this section, present-day tropospheric ozone was taken from Lamarque *et al. (*2005a); beyond 2000, tropospheric ozone was calculated by T. Wigley using the MAGICC composition model <http://www.cru.uea.ac.uk/~mikeh/software/magicc.htm> forced by the time-varying emissions of NO_x, methane and volatile organic compounds (VOCs) and these average global values were used to scale the present-day distribution. Future carbonaceous particles are scaled from their present-day distribution (Collins *et al.*, 2001) by a globally uniform factor whose time evolution follows the global evolution of SO_2 emissions. Stratospheric ozone changes are prescribed following the study by (Kiehl *et al.*, 1999). Further details on the composition models used by NCAR are found in Appendix C in the section on National Center for Atmospheric Research.

3.2.3 Tropospheric Burden

The composition models each calculate time-varying three-dimensional distributions of the short-lived gases and particles (except for CCSM where 2030 and 2050 ozone, black carbon, and organic carbon were scaled based

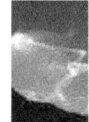

While the three models used in this study did not use identical present-day emissions, their human-caused emissions were based on the same international inventory.

Table 3.2 Global burdens. Values in parentheses are changes relative to 2000.

Species	Model	2000	2030	2050	2100
BC (Tg C)	GFDL	0.28	0.36 (29%)	0.39 (39%)	0.51
	GISS**	0.26	0.19 (-27%)	0.15 (-42%)	NA
	CCSM		(40%)	(10%)	
OC* (Tg C)	GFDL	1.35	1.59 (18%)	1.70 (26%)	2.15
	GISS	1.65	1.33 (-19%)	1.27 (-23%)	NA
	CCSM		(40%)	(10%)	
Sulfate (Tg $SO_4^=$)	GFDL	2.52	3.21 (27%)	2.48 (-2%)	1.50 (-40%)
	GISS	1.51	2.01 (33%)	1.76 (17%)	NA
	CCSM				
Dust (Tg)	GFDL	22.31	22.31	22.31	22.31
	GISS	34.84	34.84	34.84	NA
	CCSM				
Tropospheric Ozone (DU)	GFDL	34.0	38.4 (13%)	39.3 (16%)	38.2 (12%)
	GISS	31.6	41.5 (31%)	47.8 (51%)	NA
	CCSM	28.0	41.5 (48%)	43.0 (54%)	NA

*The organic carbon (OC) burdens include primary OC particles (with emissions as in Table 3.1) plus secondary OC particles (SOA). In the GFDL model, the global burden of SOA is 0.07 Tg C in this inventory. In the GISS model, organic carbon from SOA makes up ~24% of present-day OC emissions. Burdens are given in units of teragrams ($1Tg = 10^{12}$ g) for particles and in units of Dobson units (1DU globally averaged = 10.9 Tg O3) for ozone.
**GISS sulfate burdens include sulfate on dust surfaces, which makes up as much as one-half the total burden.

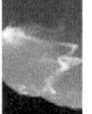

The concentrations of sulfate and carbonaceous particles are all influenced by differences in how the models simulate removal by the hydrologic cycle, accounting for at least some of the 10 to 15 percent difference in residence times in the atmosphere.

on their 2000 distributions). We compare these using the simple metric of the global mean annual average tropospheric burden (*i.e.*, the total mass in the troposphere). As was the case with emissions, the differences between the outputs of the composition models are substantial (Table 3.2). The GFDL model has a 67 percent greater present-day burden of sulfate than the GISS model, for example. As the GFDL sulfur dioxide emissions were only 13 percent greater, this suggests that either sulfate stays in the air longer in the GFDL model than in the GISS model or sulfur dioxide is converted more efficiently to sulfate in the GFDL model.

This can be tested by analyzing the atmospheric residence times of the respective models (Table 3.3). The residence time of sulfate is within ~10 percent in the two models, and in fact is slightly less in the GFDL model. This indicates that the conversion of sulfur dioxide (SO_2) to sulfate must be much more efficient in the GFDL model for it to have a sulfate burden so much larger than the GISS model. This is clearly seen in the ratio between sulfate burden and SO_2 emissions (Table 3.4). This ratio can be analyzed in terms of the total sulfur dioxide burden (in

Tg [teragrams]) per SO_2 emission (in Tg per yr); the change in SO_2 burden per SO_2 emission change, or alternatively in the percentage change in each. The latter is probably the most useful evaluation, as the fractional change will reduce differences between the starting points of the two models. We note that this metric is affected by both production and removal rates in the models. Table 3.4 shows clearly that the production of sulfate per Tg of sulfur emitted is much greater in the GFDL model than in the GISS model, either because of differences in other sources of sulfate (*e.g.*, from dimethyl sulfide [DMS]) or differences in the chemical conversion efficiency of SO_2 to sulfate (versus physical removal of SO_2 by deposition).

The residence times of black and organic carbon (BC and OC) are also fairly similar in these two models (Table 3.3). The concentrations of sulfate and carbonaceous particles are all influenced by differences in how the models simulate removal by the hydrologic cycle, accounting for at least some of the 10 to 15 percent difference in residence times. Sulfate production can vary even more from model to model, as its production from the emitted sulfur di-

Table 3.3 Global mean annual average particle residence times (days).

Species	Model	2000	2030	2050
BC	GFDL	9.4	9.4	9.3
	GISS	11.0	10.2	9.1
OC	GFDL	9.6	9.4	9.3
	GISS	8.7	8.5	8.0
Sulfate	GFDL	8.0	8.2	8.1
	GISS	8.8	8.8	9.0

oxide involves chemical oxidation, which can differ substantially between models. Removal of sulfur dioxide prior to conversion to sulfate may also be more efficient in the GISS model. In contrast, BC and OC are emitted directly, and hence any differences in how these are represented in the models would be apparent in their residence times.

The particle residence times are relatively stable in time in the GISS and GFDL models. The carbonaceous particle residence times do decrease with time in the GISS model (and to a lesser extent in the GFDL model for OC), probably owing to the shift with time from mid- to tropical latitudes, where wet and dry removal rates are different (more rapid net removal). The sulfate residence time is fairly stable over the 2000 to 2050 period. The ratio of sulfate burden to SO_2 emissions is the same for the present-day and the 2030 to 2000 changes in the GFDL model. For the 2100 to 2000 change in that model (not shown), the ratio drops from 1.00 to 0.65. As the total emissions of SO_2 decrease, a larger fraction of the sulfate production comes from DMS oxidation rather than from emitted SO_2. The conversion efficiency from SO_2 to sulfate also varies over time in the GISS model, decreasing to 2030 and increasing thereafter (inversely related to total sulfur dioxide emissions). This may reflect both

non-linearities in production (via oxidation chemistry) and the changing spatial pattern of emissions.

After comparison of the inter-model variations in particle residence times and chemical conversion efficiencies with the variations in emissions trends, it is clear that the differences in the projected changes in particle burdens in the GISS and GFDL simulations are primarily attributable to the underlying differences in emissions. This is especially true for carbonaceous particles, for which the residence times are quite similar in the models. Even though there is a greater difference in sulfate burdens due to the variations in chemical conversion efficiency between the models, the emissions trends at 2050 relative to 2000 are of opposite sign in the two models and thus dominate the difference in the burden change. Thus, the GISS model projects a greater sulfate burden at 2050 than at 2000, but substantially reduced burdens of carbonaceous particles, while the GFDL model projects the opposite, both because of the underlying emissions projections.

It is clear that the differences in the projected changes in particle burdens simulated in the two models are primarily attributable to the underlying differences in emissions.

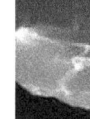

Table 3.4 Ratio of sulfate and ozone burdens to precursor emissions, global mean annual average.

Species	Model	2000 Tg burden/ Tg emission per year	2030 vs. 2000 Tg burden/ Tg emission per year	2030 vs. 2000 % burden/ % emission	2050 vs. 2000 % burden/ % emission
Sulfate	GFDL	0.017	0.017	1.00	0.08*
	GISS	0.012	0.007	0.60	0.65
Ozone	GFDL	7.19	2.24	0.32	0.44
	GISS	6.82	6.54	0.94	0.96

Ratios for sulfate are in Tg sulfate divided by Tg SO_2 per year emitted as SO_2. Ozone ratios are in Tg ozone divided by Tg N per year emitted as NO_x. Ozone values in Table 3.2 are converted to burden assuming 1 DU globally averaged = 10.9 Tg ozone. Burdens are given in units of Teragrams (Tg = 1 x 10^{12} g).
*The burden change was only 2 percent in this case, making the calculation unreliable.

Table 3.5 All-sky aerosol optical depth (550nm extinction).

Region	Particle Type	Model	2000	2030	2050	2100
Global	BC	GFDL	.0076	.0096	.0105	.0138
		GISS	.0045	.0034	.0028	NA
	Sulfate	GFDL	.1018	.1227	.0906	.0591
		GISS	.0250	.0312	.0278	NA
		CCSM	.048	.062	.052	NA
	Sea salt	GFDL	.0236	.0236	.0236	.0236
		GISS	.1065	.1080	.1050	NA
		CCSM	.018	.018	.018	NA
	Dust	GFDL	.0281	.0281	.0281	.0281
		GISS	.0372	.0389	.0387	NA
		CCSM	.0275	.0275	.0275	NA
	OC	GFDL	.0104	.0122	.0131	.0166
		GISS	.0166	.0135	.0130	NA
	Nitrate	GISS	.0054	.0057	.0060	NA
	Total	*GFDL*	*.1715*	*.1964*	*.1660*	*.1411*
		GISS	*.1959*	*.2007*	*.1934*	*NA*
		CCSM	*.116*	*.1392*	*.1206*	*NA*
Northern Hemisphere	BC	GFDL	.0109	.0147	.0161	.0209
		GISS	.0062	.0043	.0032	NA
	Sulfate	GFDL	.1509	.1766	.1038	.0694
		GISS	.0352	.0449	.0388	NA
		CCSM	.078**	.097**	.073**	NA
	Dust	GISS	.0600	.0642	.0615	NA
		GFDL	.0491	.0491	.0491	.0491
	Sea salt	GISS	.0630	.0619	.0647	NA
		GFDL	.0181	.0181	.0181	.0181
	Total	*GFDL*	*.2430*	*.2756*	*.2056*	*.1807*
		GISS	*.1910*	*.1985*	*.1907*	*NA*
		CCSM	*.1538*	*.1827*	*.1502*	*NA*
Southern Hemisphere	BC	GFDL	.0042	.0046	.0049	.0066
		GISS	.0029	.0026	.0023	NA
	Sulfate	GFDL	.0526	.0689	.0774	.0487
		GISS	.0148	.0175	.0170	NA
		CCSM	.052**	.062**	.075**	NA
	Dust	GISS	.0144	.0137	.0159	NA
		GFDL	.0071	.0071	.0071	.0071
	Sea salt	GISS	.1502	.1541	.1453	NA
		GFDL	.0291	.0291	.0291	.0291
	Total	*GFDL*	*.1000*	*.1171*	*.1263*	*.1015*
		GISS	*.1997*	*.2030*	*.1962*	*NA*
		CCSM	*.0779*	*.0957*	*.0910*	*NA*

**Total for sulfate plus sea salt.

The ozone burden in the lower atmosphere increases in the future in all three models.

The results for tropospheric ozone tell a different story. The ozone burden increases in the future in all three models, but the percentage increase relative to 2000 differs by more than a factor of three at 2030 (Table 3.2). Examining the ozone changes relative to the NO_x emissions changes, there are very large differences between the GFDL and GISS models (Table 3.4). This may reflect the influence of processes such as stratospheric ozone influx which is independent of NO_x emissions, as well as the roles of precursors such as carbon monoxide (CO) and hydrocarbons that also influence tropospheric ozone. In particular, the GISS model computed a large increase in the flux of ozone into the troposphere as the stratospheric ozone layer recovered, while the composition model used at GFDL held stratospheric ozone fixed and hence did not simulate similar large increases. In addition, there are well-known non-linearities in O_3-NO_x chemistry (Stewart *et al.*, 1977), and it has been shown that the ozone production efficiency can vary substantially with time (Lamarque *et al.*, 2005a; Shindell *et al.*, 2006a). Thus for tropospheric ozone, the differences in modeled changes of nearly a factor of three (13 *vs.* 33 percent increase) are much larger than the differences in the NO_x precursor emissions (33 *vs.* 43 percent increase).

3.2.4 Aerosol Optical Depth

The global mean present-day all-sky aerosol optical depth (AOD)[3] in the three models ranges from 0.12 to 0.20 (Table 3.5). This difference of almost a factor of two suggests that particles are contributing quite differently to the Earth's energy balance with space in these models. Observational constraints on the all-sky value are not readily available, as most of the extant measurement techniques are reliable only in clear-sky (cloud-free) conditions. Sampling clear-sky areas only, the GISS model's global total aerosol optical depth is 0.12 for 2000 (0.13 Northern Hemisphere, 0.10 Southern Hemisphere). This includes contributions from sulfate, carbonaceous, nitrate, dust, and sea-salt particles. The clear-sky observations give global mean values of ~0.135 (ground-based AERONET) or ~0.15 (satellite composites, including AVHRR or MODIS observations), though these have substantial limitations in their spatial and temporal coverage. The CCSM

---

[3] Aerosol optical depth is a measure of the fraction of radiation at a given wavelength absorbed or scattered by particles while passing through the atmosphere.

and GFDL models did not calculate clear-sky aerosol optical depth. Given that the all-sky values are larger, and substantially so in the GISS model (though this will depend upon the water uptake of particles), it seems clear that the values for CCSM would be too small compared with observations since even their all-sky values are lower than the estimate from observations. This may be related to CCSM's use of AVHRR data in assimilation of aerosol optical depth to create the CCSM climatology (Collins *et al.*, 2001, 2006), as that data appears to be low relative to MODIS observations, for example.

For all three models, there are large differences in the contributions of the various particles (Figure 3.2; Table 3.5). This is true even for GFDL and GISS models, with relatively similar all-sky global mean aerosol optical depths. More than half the aerosol optical depth in the GFDL model comes from sulfate, while this particle contributes only about one-eighth the aerosol optical depth in the GISS model. Instead, the GISS model's aerosol optical depth is dominated by the largely natural sea salt and dust particles, which together contribute 0.14 to the aerosol optical depth. These two particles contribute a much smaller aerosol optical depth in the CCSM and GFDL models, ~0.06 or less, with the differences with respect to GISS predominantly due to sea salt. The relative contribution from sulfate in the CCSM model looks similar to the GFDL model, with nearly half its aerosol optical depth coming from sulfate, but the magnitude is much smaller. It seems clear that the GFDL model's direct sulfate contribution is biased high (Ginoux *et al.*, 2006), while the GISS model's sulfate is biased low in this model version (Shindell *et al.*, 2007). However, the relative importance of the different particles is not well understood at present (Kinne *et al.*, 2006).

Large differences in the relative aerosol optical depth in the Northern Hemisphere and Southern Hemisphere are also apparent in the models (Table 3.5). The ratios of the present-day Northern Hemisphere to Southern Hemisphere total aerosol optical depths in the three models differ widely, with values of 2.43, 1.97, and 0.96 in the GFDL, CCSM, and GISS models, respectively. This clearly reflects the dominant contribution of sulfate to optical depth in the GFDL and CCSM models, as it has large anthropogenic Northern Hemisphere sources, and the dominance of sea salt in the GISS model, with its largest source being the Southern Ocean. While composite satellite data shows clearly greater aerosol optical depths in the Northern Hemisphere than the Southern Hemisphere, most satellite instruments lose coverage near the northern edge of the Southern Ocean (Kinne *et al.*, 2006). Unfortunately, quality-controlled networks such as AERONET provide virtually no ground-based data poleward of 45°S. Thus while it seems unlikely that the aerosol optical depth is larger in the Southern Hemisphere than the Northern Hemisphere, as in the GISS model, presently available data are not adequate to fully characterize this ratio, as aerosol optical depths over the Southern Ocean are poorly known.

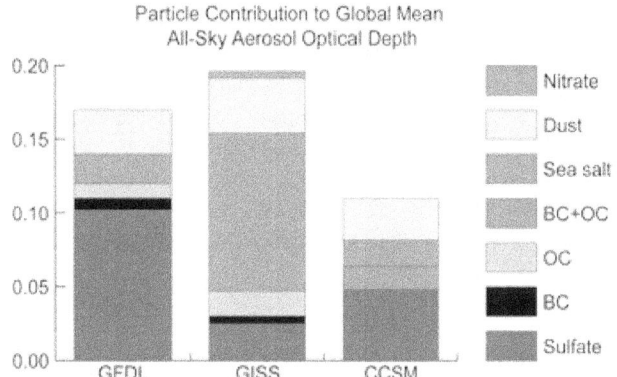

Figure 3.2 Present-day contributions from individual particles to global mean all-sky aerosol optical depth (550 nm). Neither GFDL nor CCSM include nitrate.

3.3 CLIMATE STUDIES

3.3.1 Experimental Design

The climate studies discussed here consist of transient climate simulations that were designed to isolate the climate effects of projected changes in the short-lived gases and particles and calculate their importance relative to that of the long-lived well-mixed greenhouse gases. The simulations from the GFDL, GISS, and NCAR groups each employed ensembles (multiple simulations differing only in their initial conditions) in order to reduce the unforced variability in the chaotic climate system. One three-member ensemble included the evolution of short- and long-lived gases and particles following the A1B storyline emission scenario, while the second ensemble included only the evolution of long-lived gases with the short-lived gases and particles fixed at present values. While all three groups used the same values for the long-lived gases, each had its own version of an A1B scenario for short-lived gases and particles, as discussed previously in Section 3.2.

The global three-dimensional distributions of short-lived gases and particles were modeled using each group's chemistry-particle composition model. For the first ensemble, the GFDL simulations used particle and ozone distributions computed each decade out to 2100, while the GISS and CCSM simulations employed values computed for 2000, 2030, and 2050. Either seasonally varying or monthly-average three-dimensional distributions were saved. Short-lived gas and particle concentrations for intermediate years were linearly interpolated between the values for computed years. In both sets of simulations, the concentrations of long-lived gases varied with time. In practice, NCAR performed only a single pair of simulations out to 2050, while GISS performed all three pairs out to 2050, and GFDL extended all three pairs out to 2100.

3.3.2 Climate Models
3.3.2.1 GEOPHYSICAL FLUID DYNAMICS LABORATORY (GFDL)

Climate simulations at GFDL used the comprehensive climate model (Atmosphere-Ocean General Circulation Model [AOGCM] [Box 1.1]) recently developed at NOAA's Geophysical Fluid Dynamics Laboratory, which is described in detail in Delworth *et al.* (2006). The control simulation of this AOGCM (using present-day values of radiatively active gases and particles) has a stable, realistic climate when integrated over multiple centuries. The model is able to capture the main features of the global evolution of observed surface temperature for the twentieth century as well as many continental-scale features (Knutson *et al.*, 2006). Its equilibrium climate sensitivity to a doubling of CO_2 is 3.4°C[4] (Stouffer *et al.*, 2006). The model includes the radiative effects of well-mixed gases and ozone on the climate as well as the direct effects of particles, but does not include the indirect particle effects (Box 3.1). Further details on the model resolution, model physics, and model performance are included in Appendix D in the section on Geophysical Fluid Dynamics Laboratory.

3.3.2.2 GODDARD INSTITUTE FOR SPACE STUDIES (GISS)

The GISS climate simulations were performed using GISS ModelE (Schmidt *et al.*, 2006). This model has been extensively evaluated against observations (Schmidt *et al.*, 2006), and has a climate sensitivity in accord with values inferred from paleoclimate data and similar to that of mainstream General Circulation Models; the equilibrium climate sensitivity for doubled CO_2 is 2.6°C. The radiatively active compounds in the model include well-mixed gases, ozone, and particles. The model includes a simple parameterization for the particle indirect effect (Menon *et al.*, 2002) (Box 3.1). Further details on the model resolution and model physics are included in Appendix D in the section on Goddard Institute for Space Studies.

3.3.2.3 COMMUNITY CLIMATE SYSTEM MODEL (CCSM)

The transient climate simulations use the CCSM Community Climate System Model CCSM3 (Collins *et al.*, 2006). The equilibrium climate sensitivity of this model to doubled CO_2 is 2.7°C. Further details on the model resolution

---

[4] Equilibrium climate sensitivity is defined here as the global-mean, annual-mean surface temperature change of a climate model in response to a doubling of atmospheric carbon dioxide from preindustrial levels, when the model has fully adjusted to the change in carbon dioxide.

> The model experiments used here are designed to isolate the climate effects of projected changes in the short-lived gases and particles and calculate their importance relative to that of the long-lived well-mixed greenhouse gases.

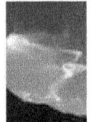

BOX 3.1: Radiative Effects of Particles

The direct effects of particles refer to their scattering and absorption of both incoming solar and outgoing terrestrial radiation. By reflecting incoming radiation back to space, most particles have a negative radiative forcing (cooling effect). For reflective particles (sulfate, organic carbon, nitrate, dust and sea salt), this effect dominates over their absorption of outgoing radiation (the greenhouse effect) on the global scale. The balance varies both geographically and seasonally as a function of solar radiation and the ground temperature. In contrast, absorbing particles such as black carbon have a positive radiative forcing (warming effect) as they absorb incoming and outgoing radiation, reducing the overall fraction of the sun's irradiance that it reflected back to space. They can also absorb outgoing radiation from the Earth (the greenhouse effect).

In addition to their direct radiative effects, particles may also lead to an indirect radiative forcing of the climate system through their effect on clouds. Two particle indirect effects are identified: The first indirect effect (also known as the cloud albedo effect) occurs when an increase in particles causes an increase in cloud droplet concentration and a decrease in droplet size for fixed liquid water content (Twomey, 1974). Having more, smaller drops increases the cloud albedo (reflectivity). The second indirect effect (also known as the cloud lifetime effect) occurs when the reduction in cloud droplet size affects the precipitation efficiency, tending to increase the liquid water content, the cloud lifetime (Albrecht, 1989), and the cloud thickness (Pincus and Baker, 1994). As the clouds last longer, this leads to an increase in cloud cover. It has been argued that empirical data suggest that the second indirect effect is the dominant process (Hansen et al., 2005).

Satellite- and ground-based observations have been used to estimate particle indirect effects and to evaluate their treatment in climate models (e.☐ Kaufmann et al., 2005; Lohmann and Lesins, 2002). The recently available CALIPSO and CloudSat measurements, which provide vertical distributions of particles and clouds, may be particularly useful for these purposes. However, the difficulty of separating the influence of particle indirect effects from dynamical and meteorological effects remains a major problem for such observational studies (Lohmann et al., 2006).

The direct effects of particles are relatively well-represented in climate models such as those described in Section 3.3.2 and used in this study, though substantial uncertainties exist regarding the optical properties of some particle types and especially of particle mixtures. Because of the inherent complexity of the particle indirect effect, climate model studies dealing with its quantification necessarily include an important level of simplification. While this represents a legitimate approach, it should be clear that the climate model estimates of the particle indirect effect are very uncertain.

The studies discussed in Chapter 3 of this Report include the direct effects of particles in all three models (though nitrate is only included in the GISS model). The indirect effect is only included in the GISS model, which uses a highly simplified representation of the second indirect effect.

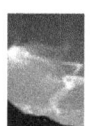

and construction are found in Appendix D in the section on National Center for Atmospheric Research.

3.3.3 Radiative Forcing Calculations

The radiative forcing at the tropopause provides a useful, though limited, indicator of the climate response to perturbations (Hansen et al., 2005) (Box 3.2).

3.3.3.1 GLOBAL AND HEMISPHERIC AVERAGE VALUES: GFDL AND GISS

Radiative forcing calculations were performed by GFDL (adjusted forcing) and GISS (instantaneous forcing), but were not performed for the CCSM model. The annual-average global-mean radiative forcing (RF) from short-lived gases and particles at 2030 relative to 2000 is small in both the GFDL and GISS models (Figure 3.3;

BOX 3.2: Radiative Forcing

Radiative forcing is defined as the change in net (down minus up) irradiance (solar plus longwave, in W per m²) at the tropopause due to a perturbation after allowing for stratospheric temperatures to adjust to radiative equilibrium, but with surface and tropospheric temperatures and state held fixed at the unperturbed values (IPCC, 2007; Ramaswamy *et al.*, 2001). This quantity is also sometimes termed adjusted radiative forcing. If the stratospheric temperatures are not allowed to adjust, the irradiance change is termed instantaneous radiative forcing.

The utility of the radiative forcing concept is that, to first order, the equilibrium global-mean, annual-mean surface temperature change is proportional to the radiative forcing, for a wide range of radiative perturbations. The proportionality constant (often denoted as the climate sensitivity parameter, λ) is approximately the same (to within 25 percent) for most drivers of climate change (IPCC, 2007), with a typical value of ~0.5-0.7 for most models. This enables a readily calculable and comparable measure of the climate response to radiative perturbations, such as those discussed in this Chapter.

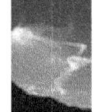

Table 3.6). However, this is for quite different reasons. In the GFDL model, a large increase in sulfate optical depth leads to a negative forcing that is largely balanced by positive forcings from increased black carbon particle and ozone. In the GISS model, increased sulfate and reduced black carbon both lead to relatively small negative forcings that largely offset a substantial positive forcing from increased ozone. Moving to 2050, the models now diverge in their net values as well as the individual contributions. The GFDL model finds a positive radiative forcing due in nearly equal parts to increased black carbon and ozone. In contrast, 2050 radiative forcing in the GISS model again reflects an offset between positive forcing from ozone and negative particle forcing, with the largest contribution to the latter from reduced levels of black carbon. Both models show a partial cancellation of the black carbon forcing by an opposing forcing from organic carbon. Thus, the two models show somewhat consistent results for ozone, but differ dramatically for black carbon and sulfate particle. By 2100, the GFDL model has a large positive radiative forcing relative to 2000, due to the continued increase in black carbon as well as the decrease in sulfate.

Inter-model differences in radiative forcing are predominantly due to differences in modeled burdens rather than to differences in the calculation of radiative properties in the models. This can be seen clearly by examining the radiative forcing-to-burden ratio, which we term the radiative efficiency (Table 3.7). This shows fairly similar values for GFDL and GISS. The largest differences are seen for black carbon, which may reflect differences in the geographic location of projected black carbon changes as well as differing treatments of the radiative properties of black carbon. Additionally, the vertical

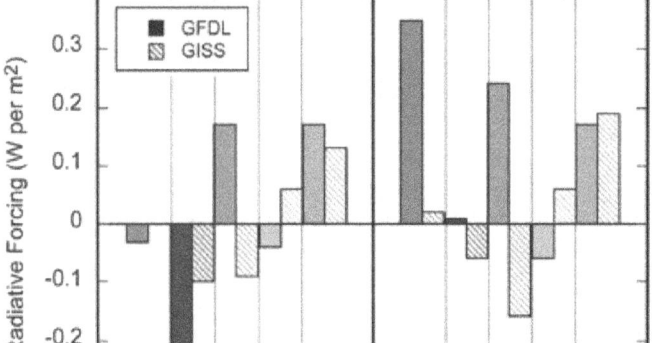

Figure 3.3 Global mean annual average radiative forcing (in W per m²) from short-lived gases and particles at 2030 and 2050 relative to 2000. Values from the GFDL model are shown as solid bars; values from the GISS model have diagonal hatching. (Note that instantaneous forcing values from the GFDL model are shown in this figure, not the adjusted forcings shown in Table 3.6.)

Table 3.6 Global mean radiative forcing for short-lived gases and particles in Watts per square meter (W per m²).

	Model	2030	2050	2100
Total	GFDL	.04	.48	1.17
	GISS	.00	.02	NA
Particles	GFDL	-.15	.24	.98
	GISS	-.13	-.17	NA
Sulfate	GFDL	-.32	.01	.51
	GISS	-.10	-.06	NA
BC	GFDL	.21	.30	.63
	GISS	-.09	-.16	NA
OC	GFDL	-.04	-.06	-.15
	GISS	.06	.06	NA
Ozone	GFDL	.19	.23	.19
	GISS	.13	.19	NA

Values are annual average radiative forcings at the tropopause (meteorological tropopause in the GISS model, "linear" tropopause in the GFDL model). "Particles" is the total of sulfate, black carbon (BC), and organic carbon (OC) (plus nitrate for GISS) particles. GISS values do not include particle indirect effects that were present in that model.

GFDL values are for adjusted radiative forcing; GISS values are for instantaneous radiative forcing (Box 3.2). The GFDL values are from Levy *et al.* (2008).

distribution of the black carbon changes will affect the radiative forcing, as will their location relative to clouds. Variations in modeling the particle uptake of water, which can have a substantial impact on the aerosol optical depth, do not seem to play a very large role in the global mean radiative forcing judging from the fairly close agreement in the two models' sulfate radiative efficiencies (Table 3.7). They may contribute to about 20 percent difference in the radiative forcing-to-burden ratios for sulfate, however. Examination of the radiative forcing-to-aerosol optical depth change (Table 3.7) shows that given a particular aerosol optical depth change, the models are in good agreement as to the resulting radiative forcing. We caution that this result contrasts with a wider model study that found larger differences in this ratio (Schulz *et al.*, 2006), though the variation in radiative forcing-to-aerosol optical depth across models was still less than the variation in aerosol optical depth itself. This suggests a possible further source of model differences that could exist were different models to be used in a study such as this.

Both the GFDL and GISS models show a positive forcing from ozone that stems partially from increased tropospheric ozone concentrations (Table 3.2) due to increased NO_x emissions (Table 3.1) and partially from the recovery of stratospheric ozone due to reductions in emis-

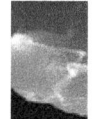

Table 3.7 Radiative efficiency.

Species	Model	(W per m²)/ Tg	(W per m²)/ AOD
BC	GFDL	2.8	104
	GISS	1.5	94
OC	GFDL	-.18	NA
	GISS	-.16	NA
Sulfate	GFDL	-.47	-16
	GISS	-.59	-16

Values are given for the radiative efficiency in terms of the radiative forcing-to-burden ratio and the radiative forcing-to-aerosol optical depth (AOD) ratio. All values are global-mean, annual-mean averages. Values for radiative forcing and burden or aerosol optical depth changes are for 2050 versus 2000 for black carbon (BC) and organic carbon (OC), and 2030 versus 2000 for sulfate in order to analyze the largest changes for each particle. GISS values for the sulfate burden changes include only the portion of sulfate not absorbed onto dust, as this portion alone is radiatively important.

sions of ozone-depleting substances (primarily halogens). The forcing from the tropospheric portion of the ozone changes is substantially more important, however (Shindell *et al.*, 2007). (The NCAR group did not calculate the radiative forcing, but forcing in their model is likely to have been similar, as they found an increase in the tropospheric ozone burden from 2000 to 2050 of 15.0 Dobson Units [DU], very close to the GISS value of 16.2 DU [Table 3.2].) As shown previously, however, the apparent sensitivity of ozone burden to changes in NO_x emissions differs substantially between the GISS and GFDL models. Thus the similarity in the radiative forcing may be largely fortuitous, resulting from a cancellation of changes in emissions and of sensitivities of ozone to NO_x emissions.

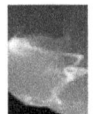

Increases in low-level ozone provide much more potential for future warming than the increase in upper-level ozone.

Thus, at 2030, differences in the physical processes in the two models dominate the differences in radiative forcing between the two models. The large divergence in radiative forcing from sulfate stems from both the chemical conversion efficiency of SO_2 to sulfate being more than a factor of two larger in the GFDL model than in the GISS model, and the greater role of sulfate in producing aerosol optical depth in the GFDL model. In addition, the GISS model includes a substantial absorption of sulfate onto dust, a process that is highly uncertain. Such a process would reduce the radiative forcing due to sulfate. At 2050, emissions and concentrations of sulfur dioxide have returned to near their 2000 level, so that these differences are not so important at this time. Hence, the 2050 differences between the two models are dominated by differences in black carbon emissions projections and not by differences in physical processes. Differences in the residence

times and radiative efficiencies for black carbon are substantial but tend to offset.

On a hemispheric scale, the GISS and GFDL models again differ greatly (Table 3.8). The GFDL model shows a very large positive forcing in the Northern Hemisphere in 2050 due primarily to reductions in emissions of sulfate precursors and increased emissions of black carbon. Increases in sulfate precursor emissions from developing countries lead to a negative forcing in the Southern Hemisphere in the GFDL model. In the GISS model, the sign of the total net forcings are reversed, with negative values in the Northern Hemisphere and positive in the Southern Hemisphere (Table 3.8). The GISS results are primarily due to the reduction in BC in the Northern Hemisphere and the influence of increased ozone and reduced OC in the Southern Hemisphere (where sea salt dominates the aerosol optical depth, so that anthropogenic particle emissions changes are relatively less important).

3.3.3.2 REGIONAL FORCING PATTERNS

The differences in hemispheric and global forcings can be attributed to strong forcing changes in particular regions, and hence to regional emissions as the radiative forcing is typically localized relatively close to the region of emissions. Comparison of the spatial patterns of radiative forcing in the GISS and GFDL models reveals that the starkest discrepancies occur in the developing nations of South and East Asia (Figure 3.4). The emissions scenario used by the GISS model projects strong increases in SO_2 emissions from India, with little change over China. In contrast, the scenario used by the GFDL model has large decreases in sulfate emissions in both regions, especially China.

The scenarios are much more similar for the developed world, with both projecting reductions in sulfate precursor emissions for North America and Europe, for example, leading to a positive radiative forcing in both cases. Differences between the scenarios are even larger for black carbon, which increases throughout most of the Northern Hemisphere in the GFDL model but decreases in the

Table 3.8 Hemispheric radiative forcing (W per m²).

	Model	2030	2050	2100
Northern Hemisphere	GFDL	.15	1.09	1.91
	GISS	-.15	-.14	NA
Southern Hemisphere	GFDL	-.09	-.14	.42
	GISS	.16	.18	NA

Values are the net annual average forcings at the tropopause in each hemisphere from particles and ozone. GISS forcing values do not include particle indirect effects that were present in that model.

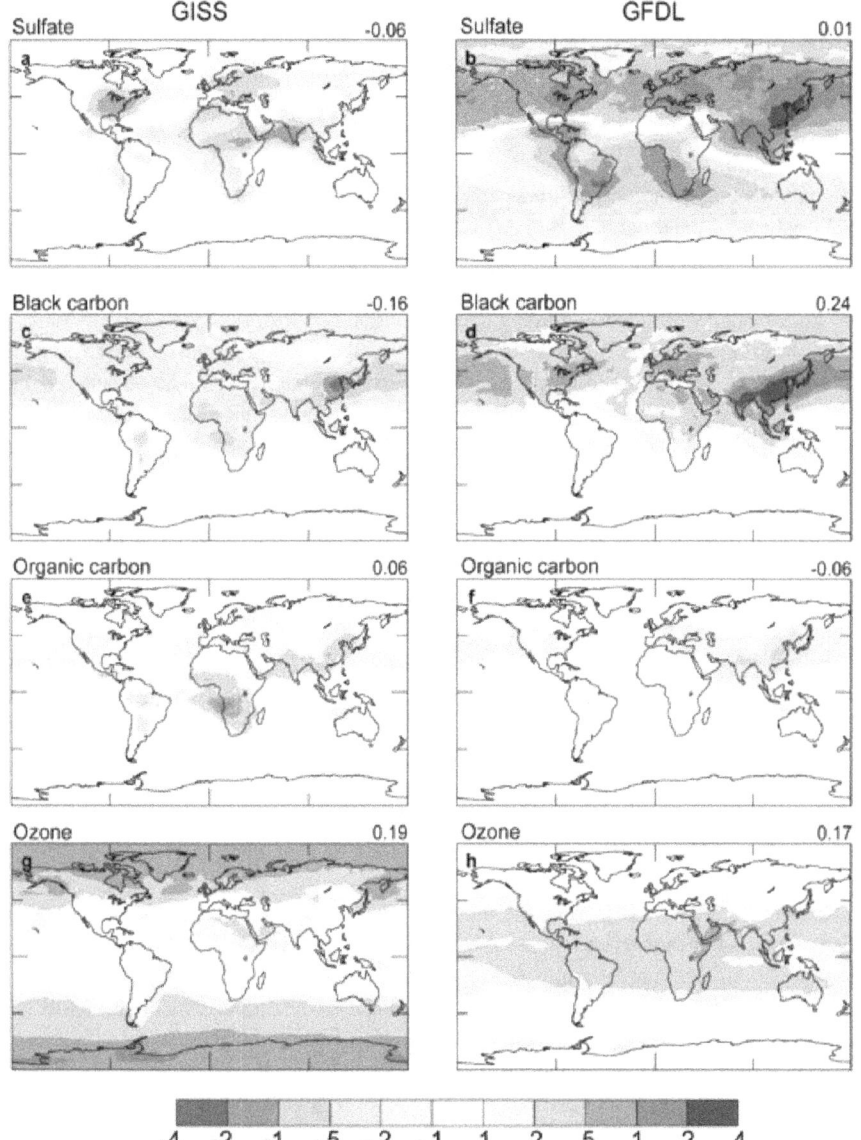

Annual Average Radiative Forcing (W per m²) Near 2050 for
Individual Short-Lived Gases and Particles

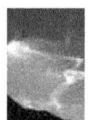

Figure 3.4 Annual average instantaneous radiative forcing (W per m²) near 2050 relative to 2000 for the indicated individual short-lived gases and particles in the GISS (left) and GFDL (right) models. Radiative forcing from long-lived gases is largely spatially uniform over the globe. (Note that the instantaneous forcings shown here for the GFDL model differ from the adjusted forcings shown in Table 3.6.)

GISS model. Again, however, the divergence is especially large over South and East Asia, where the GISS model has large reductions while the GFDL model has large increases (Figure 3.4). Thus the differences in the global total emissions discussed previously (Figure 3.1; Table 3.1) and in the global radiative forcing (Figure 3.2; Table 3.6) arise primarily from dif-

ferences in projected emissions from developing countries in Asia.

The radiative forcing from organic carbon is generally similar in its spatial pattern to black carbon, but of opposite sign and substantially reduced magnitude (25 to 40 percent of the black carbon radiative forcing). Substantial

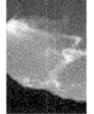

The differences in the global total emissions arise primarily from differences in projected emissions from developing countries in Asia.

differences again occur between the emissions scenarios of the two models, this time primarily over African biomass burning regions. As discussed previously, the GFDL model assumed that biomass burning emissions would scale with one-half the factor used for purely anthropogenic emissions, while the GISS model instead used regional biomass burning emissions projections (Streets *et al.*, 2004), with substantial reductions in African biomass burning.

The spatial pattern of radiative forcing from ozone is also very different in the two models (Figure 3.4). However, this forcing is not so closely tied to the region of precursor emissions in the GISS model where much of the forcing is related to an increased flux of ozone into the troposphere owing to the recovery of lower stratospheric ozone. This leads to substantial positive forcing in that model at high latitudes, even without including the effects of climate change on circulation (Section 3.3.3.4). At low latitudes, GISS shows little forcing as the modeled increase in upper stratospheric ozone causes negative radiative forcing, offsetting some of the forcing from tropospheric ozone increases, and alters lower level photochemistry. Furthermore, the particle indirect effect in that model influences cloud cover and wet deposition, which seems to reduce tropospheric ozone at low latitudes in comparison with simulations not including the particle indirect effect. The GFDL model does not show a similar high latitude enhancement, however, but instead shows maximum ozone forcing in the tropics. This may reflect a greater geographic shift in emissions to lower latitudes, a greater efficiency in transporting ozone and its precursors to the upper troposphere, where ozone has the greatest positive forcing efficiency, and differences in

the relative importance of change in the overlying stratospheric ozone column. The GFDL radiative forcing is similar to results from models with tropospheric ozone only and no particle indirect effects (Gauss *et al.*, 2003).

3.3.3.3 EFFECTS OF UNCERTAINTIES IN METHANE CONCENTRATIONS ON RADIATIVE FORCING

The SAP 3.2 simulations included methane concentrations prescribed to A1B values from the AIM integrated assessment model, for consistency with the long-lived gas runs. To investigate the potential uncertainty in the methane value derived by that integrated assessment model, the GISS model performed an additional 2050 simulation using its internal methane cycle model (Shindell *et al.*, 2007). The simulation included prescribed anthropogenic emissions increases from the AIM model to allow comparisons with the AIM results used in the results in this chapter. Natural spatially and seasonally varying emissions and soil adsorption were the standard amounts described in Shindell *et al.* (2003). Both the methane emissions from wetlands and the biogenic isoprene emissions were interactive with the climate in this run (Guenther *et al.*, 1995; Shindell *et al.*, 2004), though the distribution of vegetation did not respond to climate change.

Methane's oxidation rate is calculated by the model's chemistry scheme in both the troposphere and stratosphere. Thus methane can affect its own lifetime (which is primarily governed by tropospheric oxidation rates), as can other molecules that compete with methane for hydroxyl radicals (the main oxidizing agent), such as isoprene. The simulations included 2050 surface climate (sea surface temperatures and sea ice, taken from an earlier climate model run). Changes in water vapor induced by the altered climate affect methane oxidation in those runs. Methane was initialized with estimated 2050 abundances and the simulations were run for three years. We note that the IMAGE integrated assessment model projected a continuous increase in methane emissions; this is rather different from the increase through 2030 and slow decrease thereafter in the AIM integrated as-

sessment model. At 2050, for example, this led to projected anthropogenic methane emissions of 512 Tg C per year in the IMAGE model, substantially greater than the 452 Tg C per year from the AIM model used here (compared with 323 Tg C per year for 2000).

We find that methane emissions from wetlands increase from 195 to 241 Tg C per year while emissions of isoprene increase from 356 to 555 Tg C per year. Additionally, even in the absence of changes in emissions from natural sources, the projected anthropogenic emissions of ozone precursors (including methane itself) increase the lifetime of methane while climate change reduces it via increased temperature and water vapor (Table 3.9). These responses to anthropogenic emissions and to climate change without interactive emissions are qualitatively consistent with those reported from a range of models (using different emissions projections) in Stevenson *et al.* (2006). The effect of precursor emissions is stronger in our scenario, so that the net effect of anthropogenic emissions and climate changes is to increase methane's lifetime. When natural emissions are also allowed to respond to climate change, increased competition from isoprene and increased methane emissions from wetlands lead to further increases in methane's lifetime (Table 3.9) and enhanced methane abundance.

The 2050 simulation with the model's internal methane cycle had a global mean surface methane value of 2.86 ppmv in year three, with sources exceeding sinks by 80 Tg C per year (a growth rate that may reflect an overestimate of the loss rate in the AIM model used in the initial guess). Extrapolating the change in methane out to equilibrium using an exponential fit to the three years of model results yields a 2050 value of 3.21 ppmv.

We have calculated radiative forcings using the standard calculation (Table 6.2 in Ramaswamy *et al.*, 2001) assuming an increase in N_2O from 316 to 350 ppb in 2050, following the A1B "marker" scenario (using the AIM integrated assessment model). The 2050 methane forcing using the methane concentration specified in the A1B "marker" scenario would be 0.22 W per m^2 while using the larger methane concentrations of 2.86 or 3.21 ppmv calculated with our

model gives 0.36 or 0.46 W per m^2, respectively. Of course, it is difficult to estimate methane's abundance at a particular time without performing a full transient methane simulation. However, uncertainty in the forcing from methane appears to be at least 0.1 to 0.2 W per m^2. Note that use of the 40 percent larger anthropogenic methane emissions increase from the IMAGE integrated assessment model would have led to a substantially larger forcing. Should the results of our modeling of the methane cycle prove to be robust, this would imply that future positive forcing from methane might be substantially larger than current estimates based on integrated assessment model projections.

We note that while the A1B projections assume a substantial increase in atmospheric methane in the future, the growth rate of methane has in fact decreased markedly since the early 1990s and leveled off since ~1999 (Dlugokencky *et al.*, 2003). Hence, the projections may overestimate future atmospheric concentrations. However, there are indications that the growth rate decrease was primarily due to reduced anthropogenic emissions, and that these have been increasing again since 1999 (though masked by a coincident decrease in natural methane emissions) (Bousquet *et al.*, 2006). All of this suggests that atmospheric methane may in fact increase substantially again in the future, as assumed by the integrated assessment models, although other methane studies have argued for an increase in its principal loss path as the explanation, rather than changes in emissions (*e.g.*, Fiore *et al.*, 2006). Other emissions, such as NO_x from lightning and from soil and dimethyl-sulfide from the oceans, are also expected to respond to climate change. Changes in land

There is considerable uncertainty in methane-driven climate changes in future scenarios. They could be larger than currently anticipated.

Table 3.9 Methane lifetime in GISS simulations. Includes calculated photochemical loss (in troposphere and stratosphere) and prescribed 30 Tg C per year loss to soils.

Run	Lifetime (years)
2000	9.01
2030	9.96
2050	10.39
2030 with climate change	9.72
2050 with climate change	10.01
2050 with methane cycle	10.42

cover would also affect both emissions and removal of trace gases and particles. Further work is required to gauge the importance of these and other climate-chemistry feedbacks.

3.3.3.4 EFFECTS OF CLIMATE CHANGE ON RADIATIVE FORCING

The chemical composition simulations (Section 3.2) did not include the effects of climate change on the short-lived gases and particles, only the effects of projected changes in anthropogenic emissions. Separate sets of simulations with the GISS model included climate change via prescribed sea-surface temperatures and sea-ice cover taken from prior runs. Climate change increased the radiative forcing from ozone by increasing stratosphere-troposphere exchange (STE) and hence ozone near the tropopause where it is most important radiatively (Hansen *et al.*, 1997). This effect outweighed increased reaction of excited atomic oxygen with the enhanced tropospheric water vapor found in a warmer climate, which led to ozone reductions in the tropical lower troposphere. The overall impact was to increase radiative forcing by .07 W per m² in 2050. Climate change slightly increased the negative forcing from sulfate (by .01 W per m²), consistent with an increase in tropospheric ozone in these runs (as ozone aids in sulfur dioxide oxidation both directly and via hydroxyl formation).

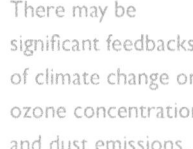

Dust emissions decreased slightly (about five percent at 2050) in these climate runs, but there was more sulfate on dust, suggesting that this played only a minor role in the sulfate forcing response to climate change. The reduction in dust would itself lead to a slight negative forcing (approximately 0.02 W per m²). However, emissions in the model respond only to changes in surface wind speeds, and not to changes in sources due to either CO_2 fertilization or climate-induced vegetation changes which have a very uncertain effect on future dust emissions (Mahowald and Luo, 2003; Woodward *et al.*, 2005).

Much of the increase in ozone forcing results from an increase in stratosphere-troposphere exchange in the GISS model of 134 Tg per year (about 20 percent of its present-day value) as climate warms. An increase in transport rates between the stratosphere and the troposphere is

a robust projection of climate models (Butchart *et al.*, 2006). Combined with the expected recovery of stratospheric ozone, this should enhance the influx of stratospheric ozone into the lower atmosphere. However, the net effect of climate change on ozone is more difficult to determine as it results from the difference between enhanced stratosphere-troposphere exchange and enhanced chemical loss in the troposphere in a more humid environment, which is not consistent among climate models (Stevenson *et al.*, 2006).

3.3.4 Climate Model Simulations 2000 to 2050

As discussed in Section 3.3.1, the experimental design consists of two sets of simulations: (1) the effects of changes in short-lived and long-lived gases and particles in the twenty-first century (employing the A1B scenario for the evolution of the long-lived gases and the output from the composition models discussed in Section 3.2 for the short-lived gases and particles); (2) the effects of changes in the long-lived gases only, with the short-lived gases and particles concentrations held at 2000 values. The effects of short-lived gas and particle changes are determined by subtracting the climate responses of the runs with changes in long-lived gases only from those with changes in long-lived and short-lived gases and particles. This procedure is justified by studies showing that the climate response to changes in radiatively active species is generally linear (Ramaswamy and Chen, 1997; Haywood *et al.*, 1997; Cox *et al.*, 1995). Simulations where only the short-lived gases and particles change were not included in the experimental design, because such scenarios are neither realistic nor policy relevant.

3.3.4.1 SURFACE TEMPERATURE CHANGES

The global-mean annual-mean surface temperature responses to short-lived gases and particles in the three different models are not as dissimilar as one might have expected, given the different emissions used and the different physical processes included. The CCSM model ran only a single simulation, which showed little or no statistically significant[5] effects of the short-lived gases and particles on global mean surface temperatures. The GFDL and

There may be significant feedbacks of climate change on ozone concentration and dust emissions.

---

[5] The statistical methods used to assess significance are discussed in Box 3.3.

BOX 3.3: Statistical Methods

A result is deemed to be statistically significant if it is unlikely to have occurred by chance (*i.e.*, the probability that it occurred by chance is less than some specified threshold). A 95 percent confidence level means that the odds are 20:1 against the result having occurred by chance.

Statistical significance in the GFDL climate model results was evaluated using two approaches. For global-mean, or hemispheric-mean results involving a temperature departure from the initial (2000) value, the range (highest to lowest temperature change) of the three ensemble members used to obtain the ensemble-mean result was computed. The ensemble-mean result was deemed significant if that range was entirely different from zero. For regional (latitude-longitude) results comprising the difference of two time series, as in the evolution of temperature change due to short-lived species, the Student's-*t* test for significance was applied at each model grid point, with the result deemed significant if the statistical test showed significance at the 95 percent confidence level.

GISS models both ran three-member ensemble simulations, and both show a statistically significant[5] warming effect from short-lived gases and particles from around 2030 to the end of the runs (Figure 3.5). The GFDL model shows a warming of 0.28°C (ensemble mean 2046 to 2050). This value is commensurate with the adjusted radiative forcing of about 0.48 W per m^2 computed for 2050. The GISS model shows substantially more warming (approximately 0.13°C near 2050) than would be expected from the direct radiative forcing in that model and its climate sensitivity (λ = approximately 0.6°C [W per m^2 per year]) owing to the presence of the particle indirect effect, which contributes additional warming as particle loading decreases in the future (Shindell *et al.*, 2007).

Changing levels of short-lived gases and particles contribute approximately 20 percent of the overall global annual average warming in these two models (17 percent for GISS and 27 percent for GFDL based on 2046 to 2050 *vs.* the first five years of the run). It is important to note, however, that these models respond as they do for different reasons.

In the GFDL simulations, reduced sulfate and increased black carbon and ozone all combine to cause warming. In contrast, in the GISS model, the warming results from increased ozone and a reduced particle indirect effect, with a substantial offset (cooling) from reduced black carbon. The lack of a substantial effect from short-lived gases and particles in the CCSM simulations is attributable to the emissions used, which produce small increases in

sulfate (cooling) and small increases in black carbon (warming) that largely offset one another (thus their radiative forcing changes little from 2000 to 2050).

Hemispheric temperatures show trends largely consistent with the radiative forcings (Table 3.8), namely substantial warming in the Northern Hemisphere in the GFDL model and in the Southern Hemisphere in the GISS model (Figure 3.6). The Northern Hemisphere warming in the GFDL model is driven primarily by the large decreases projected for sulfate and the large increase projected for black carbon in that model for the industrialized areas of the

Two of the three models show substantially increased global warming as a result of changes in short-lived gases and particles.

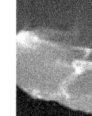

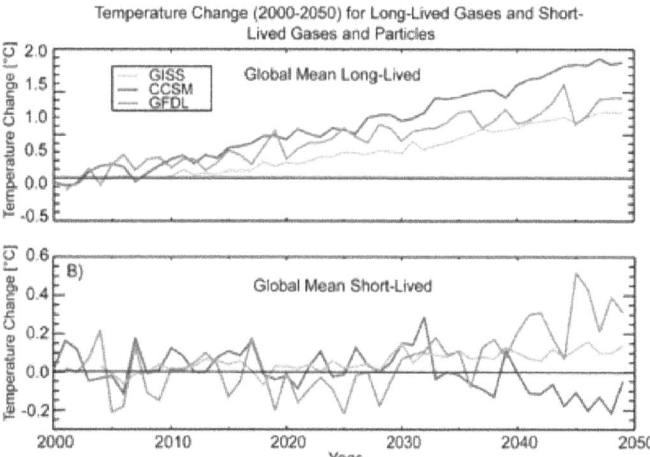

Figure 3.5 Global mean annual average temperature change (°C) in the simulations with time-varying long-lived (top) and short-lived (bottom) gases and aerosols. Results are three-member ensemble means for GFDL and GISS and single-member simulations for CCSM. Results for the short-lived gases and aerosols are obtained by subtraction of the (long-lived) calculations from the (short + long-lived) calculations.

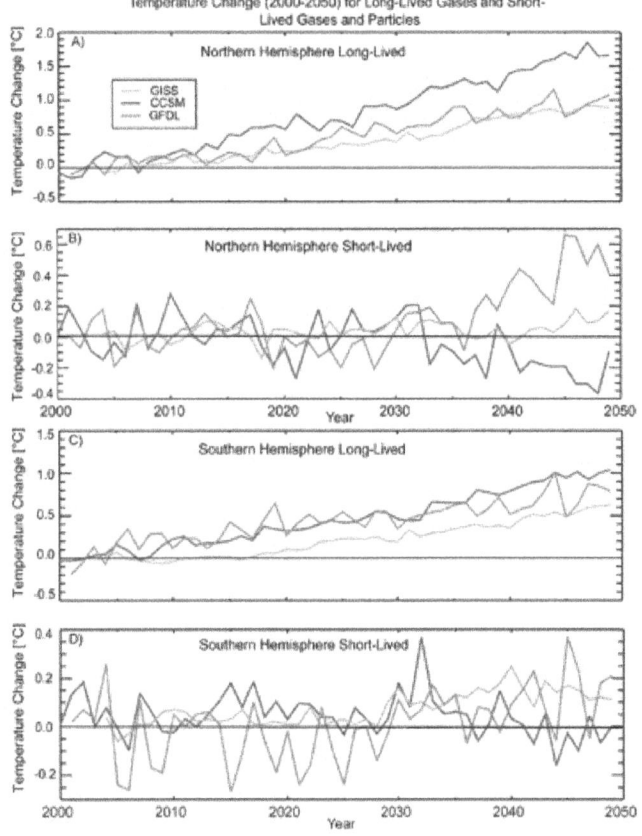

Temperature Change (2000-2050) for Long-Lived Gases and Short-Lived Gases and Particles

Figure 3.6 Hemispheric mean annual average temperature change (°C) in the simulations with time-varying long-lived and short-lived gases and particles. All the results are three-member ensemble means for GFDL and GISS and single member simulations for CCSM. Results for the short-lived gases and particles are obtained by subtraction of the (long-lived) calculations from the (short+long-lived) calculations.

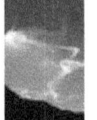

Northern Hemisphere (Levy *et al.*, 2008). This causes the aerosol optical depth from sulfate to drop by one-third in the Northern Hemisphere by 2050 while the aerosol optical depth from black carbon increases by 50 percent. The large change in sulfate dominates the overall aerosol optical depth change in that model (Table 3.5). The magnitude of the Northern Hemisphere warming is ~0.5°C by 2050, consistent with the ~1.1 W per m² radiative forcing in that model, when one accounts for the fact that the warming has not been fully realized due to the lag-time for oceanic heat adjustment (Stouffer, 2006). There is an overall negative forcing in the Southern Hemisphere in the GFDL model, as sulfate precursor emissions increase in the developing world while black carbon changes little. Some of the negative forcing from particles in the Southern Hemisphere is offset by

positive forcing from ozone, which increases rather uniformly over much of the world in that model (Levy *et al.*, 2008), leading to a small net effect and minimal temperature change from short-lived gases and particles (Figure 3.6).

The change in the forcing due to the particle indirect effect in the GISS model was argued to be on the order of 0.1 W per m² in 2050 (Shindell *et al.*, 2007). Combining this with the GISS hemispheric radiative forcings (excluding the indirect effect) in Table 3.8 yields a Northern Hemisphere radiative forcing near zero and a Southern Hemisphere forcing of about 0.3 W per m². These forcings are consistent with the warming of about 0.15°C seen in that model in the Southern Hemisphere and the lack of response in the Northern Hemisphere. Northern Hemisphere aerosol optical depth changes are dominated by a substantial reduction in black and organic carbon (the black carbon aerosol optical depth in the Northern Hemisphere falls by nearly 50 percent), which more than offsets a slight increase in sulfate (particularly as this model is less sensitive to sulfate). These particle changes lead to negative Northern Hemisphere forcing. In the Southern Hemisphere, the GISS model shows only small changes in particles, so that positive forcing from ozone dominates the net radiative forcing. The particle indirect effect further accentuates the positive forcing owing to reductions in black carbon and organic carbon. The signs of the temperature response in the two hemispheres are thus opposite in the GISS model to what they are in the GFDL model.

As for the global case, trends in the CCSM model are not significantly different in the runs with and without short-lived gases and particles. This is the result of only a miniscule change in aerosol optical depth in the Northern Hemisphere (-2 percent), as sulfate and carbonaceous particle precursor emissions are both near their present-day values by 2050 in that model. In the Southern Hemisphere, there is an increase in aerosol optical depth from 2000 to 2050, which seems to be primarily due to sulfate, but this is largely offset by increased ozone in the Southern Hemisphere as stratospheric ozone recovers.

Thus it is clear that at global and especially at hemispheric scales, the three climate models are being driven by substantially different trends in their short-lived gases and particles. These differences in particles are largely related to the differences in the projected emissions of particle precursors, though there is some contribution from differences in particle modeling as discussed previously. Additionally, the climate response of each model is different to some extent owing to the inclusion of different physical processes in the models, especially the inclusion of the particle indirect effect in the GISS model. However, the above analysis strongly suggests that the largest contributor to the inter-model variations in projected warming arise from different assumptions about emissions trends.

At smaller spatial scales, the annual average patterns of surface temperature changes induced by the short-lived gases and particles show even larger divergences (Figure 3.7).

Around 2030, the largest responses are seen at Northern middle and high latitudes. These show large regions of both cooling and warming that are characteristic of the response to changes in atmospheric circulation. Most of the response at middle and high latitudes is not statistically significant in the models owing to large natural variability. Surprisingly, all three models show similar patterns of cooling near Alaska and a region of warming over Siberia. Other regions, such as the Labrador Sea/Baffin Island area or Scandinavia, show substantial variations between models, again suggesting these middle and high latitude dynamic responses are not robust.

In the tropics, where dynamic variability is much smaller, the models find much greater areas with statistically significant responses, especially by 2050. The CCSM model finds a small but significant cooling over tropical oceans, while the other models find warming.

It is clear that especially at hemispheric scales, the three climate models are being driven by substantially different trends in short-lived gases and particles.

Pattern of Surface Temperature Change for Short-Lived Gases and Particles and Long-Lived Gases

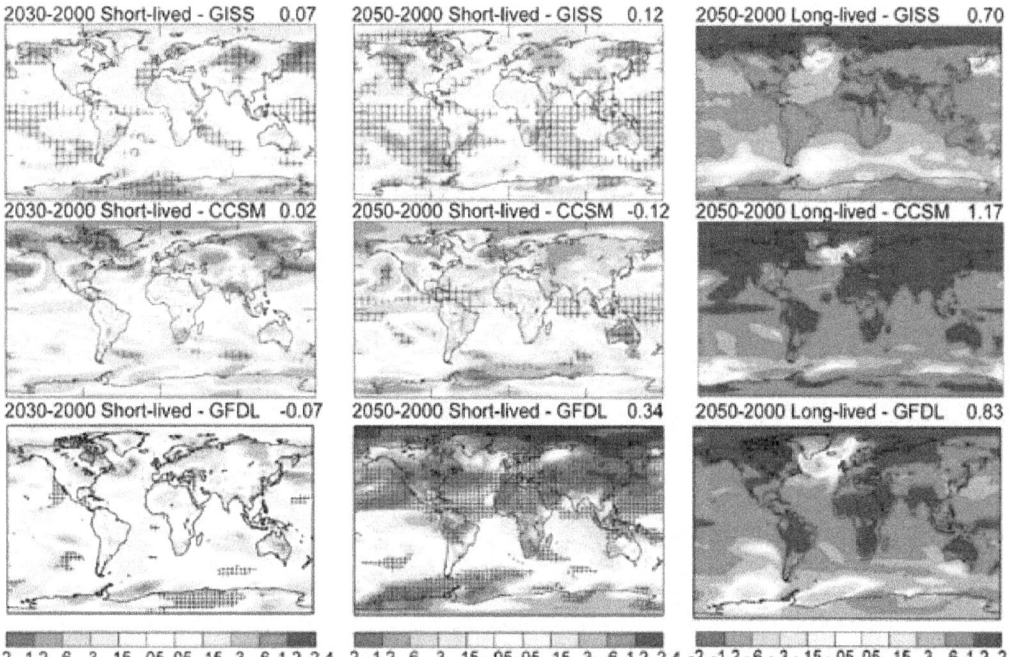

Figure 3.7 Annual average surface temperature response (°C) in the climate models to short-lived gases and particles (left and center columns) and long-lived gases (right column) changes for the indicated times. The changes at 2030 are 2020 to 2029 in the CCSM and GFDL models and 2028 to 2033 in the GISS model. At 2050, they are 2040 to 2049 in the CCSM model, 2046 to 2055 in the GFDL model, and 2040 to 2050 in the GISS model. Hatching indicates statistical significance (95 percent) for the response to short-lived gases and particles. All colored values above 0.1°C are statistically significant in the response to long-lived gases. Values in the upper right corners give the global mean.

As in the global-mean case, this appears to arise from differences in particle burdens and aerosol optical depths.

In the Arctic, the GISS and CCSM models find primarily a cooling effect from projected changes in short-lived gases and particles (especially near 2030 for GISS, and 2050 for CCSM). In contrast, the GFDL model finds a substantial warming there. This may be due in part to the increasing trend in black carbon in that model.

In the Antarctic, the GISS model shows warming related primarily to stratospheric ozone recovery. The GFDL model shows a similar result by 2050 (after which stratospheric ozone was unchanged in that model). CCSM does not show as clear an Antarctic warming, however, even though this model also included recovery of ozone in the Antarctic lower stratosphere. This is surprising given that the CCSM model appeared to show a substantial response to ozone depletion in analyses of the Southern Hemisphere circulation in IPCC AR4 simulations (Miller *et al.*, 2006b). That analysis showed that most climate models found a general strengthening of the westerly flow in the Southern Hemisphere in response to stratospheric ozone depletion. A stronger flow isolates the polar region from lower latitude air, leading to cooling over the Antarctic interior and warming at the peninsula. Conversely, recovery should lead to warming of the interior (enhanced by the direct positive radiative forcing from increased ozone), as in the GISS and GFDL simulations. However, in the GISS model the effect diminishes with time, suggesting that other aspects of

The United States and other Northern Hemisphere industrialized regions might be especially sensitive due to the projected reduction in sulfate precursor emissions in the Northern Hemisphere.

the response to short-lived gases and particles become more important in these scenarios over time, presumably as projected particle changes grow ever larger.

Warming over the central United States is present in the GISS model at all times (but is not statistically significant), in the GFDL model from about the 2040s on, and in the CCSM model around 2030, but not at 2050. The United States and other Northern Hemisphere industrialized regions might be especially sensitive to the projected reduction in sulfate precursor emissions in the Northern Hemisphere. This effect is especially large in the GFDL model, where forcings from sulfate decreases and black carbon increases both contribute to warming, though it should be noted that the largest radiative forcing is over Asia, not over the United States and Europe (Figure 3.4). In the CCSM model, the warming effect vanishes by 2050 as both sulfate and black carbon decrease, producing temperature responses that cancel. In the GISS model, reductions in sulfate and increases in ozone both contribute to warming; however, these are partially offset by cooling from reduced black carbon.

The surface temperature changes induced by the long-lived gases are clearly much larger than those induced by short-lived gases and particles over most of the Earth by 2050 (Figure 3.7). In some regions, however, the two are of comparable magnitude (*e.g.*, the polar regions and parts of the Northern midlatitude continents in the GFDL model, parts of the Southern Ocean in the GISS model) though the statistical significance of the signal for short-lived gases and particles is marginal. Consistency between the models is also clearly greater in their response to long-lived than to short-lived gases and particles.

Overall, it is clear that the regional surface temperature response does not closely follow the regional radiative forcing patterns based on either GISS or GFDL results. Both models show very large forcings over East Asia, for example, yet have minimal response there. This is especially clear when comparing the seasonal radiative forcings and climate response (Figure 3.8). Though some of the spatial mismatches could result from a lag in the climate response to seasonally varying forcings, the divergence between the patterns of forcing and response is large even for areas with minimal seasonality in the forcing (*e.g.*, Africa, subtropical Asia). See Section 9.2.2.1 of Chapter 9 in the Fourth IPCC assessment (Hegerl *et al.*, 2007) and references therein for further discussion of this issue.

3.3.4.2 PRECIPITATION, SEA-LEVEL, AND OTHER VARIABLES.

Changes in other climate variables (such as precipitation and sea level) due to short-lived gases and particles are typically too small to isolate statistically. In the case of precipitation, this is due to the high variability of the precipitation signal. The relatively short (50 year) length of the integrations discussed here accounts for the lack of a significant sea-level signal. Sea level is expected to rise in a warming climate, chiefly due to thermal expansion of the oceans. In the GISS and GFDL models, the thermal expansion would be enhanced by about 20 to 25 percent due to changes in short-lived gases and particles. Similarly, the enhancement of precipitation along the equator and drying of the subtropics that is a robust feature of climate models in a warming climate (Held and Soden, 2006) would also be accentuated in the GFDL and GISS models with their significant tropical warming, though probably not under the CCSM scenario. Such a feature can indeed be seen in the GISS response in the Atlantic and Indian Oceans.

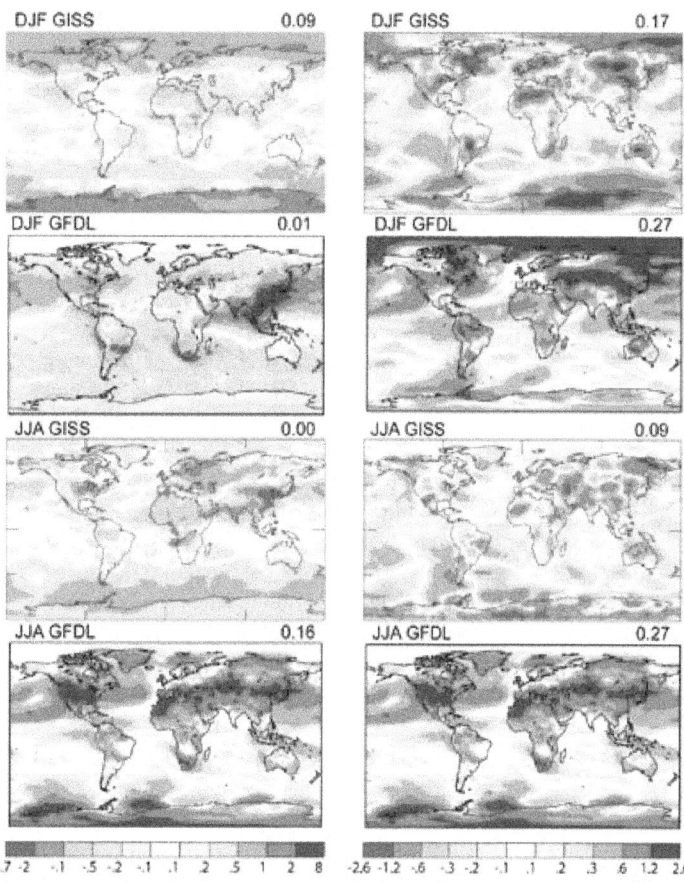

Seasonal Average Instantaneous Radiative Forcing in 2050 from
Short-Lived Gases and Particles and Surface Temperature

Figure 3.8 Seasonal average instantaneous radiative forcing (W per m²) in 2050 from short-lived gases and particles (left column) and surface temperature response in °C (right column) in the GISS and GFDL models. Boreal winter (December through February) is shown in the top two rows, while boreal summer (June through August) is shown in the bottom two rows. The temperature changes at 2050 are 2046 to 2055 in the GFDL model and 2041 to 2050 in the GISS model. Values in the upper right corners give the global mean. (Note that the instantaneous forcings shown here for the GFDL model differ from the adjusted forcings show in Table 3.6.)

On a regional scale, there are some suggestions of precipitation trends but statistical significance is marginal for either annual or seasonal changes. The CCSM model shows reductions in winter precipitation due to short-lived gases and particles across most of the United States in the 2040s, and reductions in summer precipitation in the southeastern part of the United States. That model also suggests an increase in summer monsoon rainfall over South Asia. In contrast, the GISS model shows slight increases in winter precipitation over the central United States, and a mixed signal in summer (and spring) with increased precipitation over the

Southeast and Southwest United States, but decreases over the Northeast United States. During fall, precipitation decreases over most of the country. As in the CCSM model, there is an increase in summer (and fall) precipitation over South Asia. In the annual average, the GFDL model shows no statistically significant trend over the United States. Given that significant trends are hard to identify in any of the models, and that the models do not agree on the trends themselves, we believe that it is not possible to reliably estimate precipitation trends owing to short-lived gas and particle changes under the A1B storyline.

The surface temperature changes induced by the long-lived gases are clearly much larger than those induced by short-lived gases and particles over most of the Earth by 2050 (Figure 3.7). In some regions, however, the two are of comparable magnitude.

49

3.3.4.3 DISCUSSION

In the transient climate simulations, three climate models examined the response to projected changes in short-lived gases and particles. The results shown in Figure 3.7 differ substantially among the models, particularly by 2050. Comparison has shown that the differences in the underlying emissions projections are the dominant source of inter-model variations in projected particle trends. These variations result from differences between the various integrated assessment models that provided those projections and to assumptions made about emissions not provided by the integrated assessment models.

Uncertainties in the representation of radiative processes are an additional source of differences in the temperature response of the three models. For example, the GFDL model's aerosol optical depth is substantially more sensitive to sulfate than the GISS model, with the CCSM model in between. This is partially due to the inclusion of sulfate absorption onto dust being present only in the GISS model. Additionally, the indirect effect of particles is included only in the GISS model. Thus, the inclusion of different physical processes plays a role in the inter-model differences, and is especially important near 2030, when sulfur dioxide (SO_2) emissions are near their peak. With the inclusion of the particle indirect effect, the GFDL model might yield a substantially larger warming, given that sulfate is the largest contributor to particle mass globally and that the GFDL sulfate concentrations decrease beyond 2030.

Differences in the models' representation of the hydrologic cycle, which removes soluble gases and particles, and in oxidation may also produce variations in the temperature response. Inter-model differences between the GFDL and GISS models in the residence times of particles are substantial for sulfate, and differences in the radiative effect of black carbon are also potentially sizeable. In nearly all cases, however, these are outweighed by emissions differences. Exceptions are sulfate in 2030 and tropospheric ozone; differences in the modeled conversion of SO_2 to sulfate and the sensitivity of ozone to NO_x emissions are larger than differences in projected precursor emissions. Hence, in these cases uncertainties in physical

processes, including chemistry, dominate over uncertainties in emissions. These differences between GFDL and GISS model results appear to be representative of other intermodel differences, such as those identified for modeling of sulfate and black carbon particles in recent comparisons between a large suite of models (Schulz *et al.*, 2006).

We also reiterate that uncertainties in the particle indirect effect and in internal mixing between particle types are either not included at all or only partially in these simulations. Sensitivity studies and analysis of the GISS model results indicates that the forcing from reductions in the particle indirect effect is roughly 0.1 to 0.2 W per m², while the inclusion of sulfate absorption onto dust reduces the negative forcing from sulfate at 2030 or 2050 by up to 0.2 W per m² (Shindell *et al.*, 2007), These sensitivities suggest that uncertainties in these processes could alter the global mean projected temperature trends by up to 0.1°C at 2030 or 2050, a value comparable to the total temperature trend in that model. Hence without the particle indirect effect, the GISS model would likely have shown minimal warming, while without sulfate absorption onto dust surfaces, it would likely have shown a substantially greater warming trend (at least at 2030).

The responses of methane (and other hydrocarbon) emissions and of stratosphere-troposphere exchange to climate change can also potentially have significant impacts on radiative forcing, and these processes were not included in these simulations. As discussed in sections 3.3.3.3 and 3.3.3.4, the resulting changes to radiative forcing could again substantially alter the projected temperature trends. Additionally, given the large influence of uncertainties in emissions projections, we stress that the magnitude and even the sign of the effects of short-lived gases and particles on climate might be different were alternative emissions projections used in these same models. Thus, the response of short-lived gases and particles and methane to emissions changes and climate changes has been only partially characterized by the present study, and substantial work remains to reduce uncertainties and further clarify their potential role in future climate change.

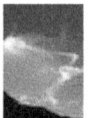

The results clearly indicate that the spatial distribution of radiative forcing is generally less important than the spatial distribution of climate response in predicting climate impact. Thus, both short-lived and long-lived gases and particles appear to cause enhanced climate responses in the same regions of high sensitivity rather than short-lived gases and particles having an enhanced effect primarily in or near polluted areas. This result is supported by analysis of the response to larger radiative perturbations in these models for the future (Levy *et al.*, 2008) and the past (Shindell *et al.*, 2007). It is also consistent with earlier modeling studies examining the response to different inhomogeneous forcings than those investigated here (Mitchell *et al.*, 1995; Boer and Yu, 2003; Berntsen *et al.*, 2005; Hansen *et al.*, 2005). This suggests that the mismatch between model simulations of the regional patterns of twentieth century climate trends and observations is likely not attributable to unrealistic spatially inhomogeneous forcings imposed in those models. Instead, the models may exhibit regional climate sensitivities that do not match the real world, and/or some of the observed regional changes may have been unforced (*i.e.*, the result of internal variability [see Knutson *et al.*, 2006 for a discussion of this issue]).

3.3.5 Climate Simulations Extended to 2100

Following the A1B "marker" scenario into the second half of the 21st century for both three-member ensembles, the GFDL simulations (Levy *et al.*, 2008) find significant climate impacts due to emissions of sulfur dioxide (SO_2), the precursor of sulfate particle, (which decrease to ~35 percent of 2000 levels by year 2100) and of black carbon (scaled to carbon monoxide emissions in the GFDL model) which continue to increase. This is confirmed by their respective radiative forcing values for 2100 in Table 3.6. By 2080 to 2100, these projected changes in emissions levels of short-lived gases and particles contribute a significant portion of the total predicted surface temperature warming for the full A1B scenario; 0.2°C in the Southern Hemisphere, 0.4°C globally, and 0.6°C in the Northern Hemisphere as shown in the time series of yearly average surface temperature change in Figure 3.9.

In Figure 3.10 we examine Northern Hemisphere summer surface temperature change between the 2090s (average over 2091 to 2100) and the 2000s (average over 2001 to 2010) due to the changes in emissions of short-lived gases and particles between the first decade and the last decade of the 21st century. Note the large warming in the Northern Hemisphere mid-latitudes with the major hot spots over the continental United States, Southern Europe and the Mediterranean. Eastern Asia, the region with the strongest radiative forcing due to changes in emissions and loading of short-lived gases and particles, is not one of the hot spots. The mid-latitude warming belt is statistically significant to the 95th percentile, with some regions significant to the 99th percentile. By contrast, the annual-average and seasonal-average patterns of change in precipitation due to changes in emissions of short-lived gases and particles (not shown) are, in general, not statistically significant.

We now focus on the large summertime warming over the United States and consider the twenty-first century time series shown in Figure 3.11. By 2100, the change in short-lived gases and particles, primarily the decrease in sulfate and increase in black carbon particles over Asia, contributes ~1.5°C of the total temperature warming of ~4°C predicted for the summertime continental United States when the effects of changes in both short-lived and long-lived gases and particles are included.

The geographic distribution of radiative forcing is generally less important than the spatial distribution of climate response in predicting climate impact.

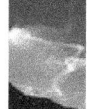

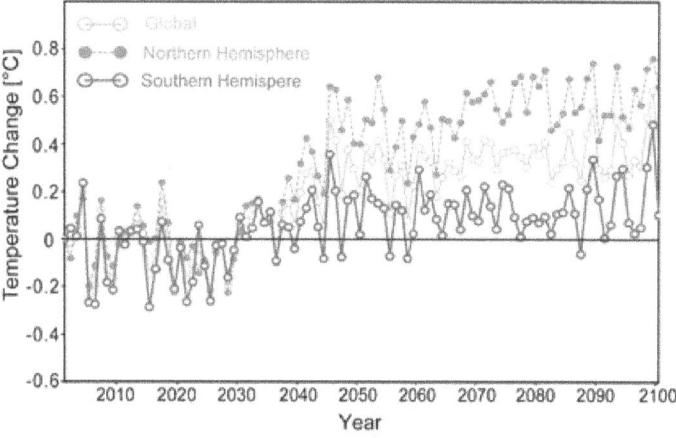

Figure 3.9 Surface temperature change in °C (2000 to 2100) due to short-lived gases and particles in the GFDL model.

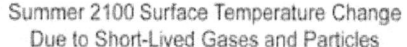
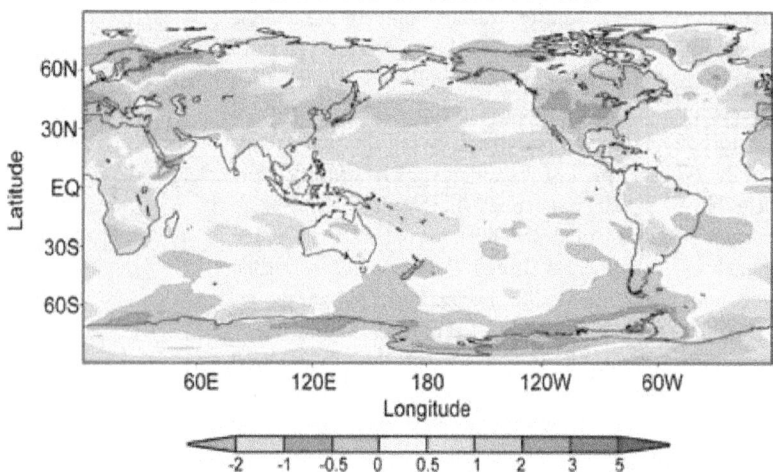

Figure 3.10 Surface temperature change in °C due to short-lived gases and particles during Northern Hemisphere summer for 2100 to 2091 vs. 2010 to 2001 in the GFDL model.

Model simulations indicate that short-lived gases and particles could account for a significant portion of the warming in 2100.

In Figure 3.12, we focus more narrowly on the central United States where the strong summertime warming was predicted for 2100. Monthly-mean area-averaged values of temperature, precipitation, and available root-zone soil water are shown for both the full A1B emissions scenario (dashed lines), where both long-lived greenhouse gases and short-lived gases and particles change and the emissions scenario with only long-lived greenhouse gases changing and short-lived gases and particle levels fixed at 2001 values (solid lines). The values are ensemble averages over the last 40 years (2061 to 2100) for each simulation. Here the climate model does predict a statistically significant (at the 95 percent confidence level) decrease in precipitation due to the change in

short-lived gases and particles (blue curves in Figure 3.12 and green curve in Figure ES.1). We next consider root-zone soil water, a quantity that integrates and responds to both temperature and precipitation. There is a statistically significant (at the 95 percent confidence level) decrease of up to 50 percent in available root-zone soil water in the Central United States during late summer (July through September), which could have important consequences for United States grain production, and merits future attention. This is the result of a global increase in short-lived gas and particle forcing, located primarily over Asia, which in turn results from the large changes projected by the A1B "marker" scenario for Asian emissions of SO_2 and black carbon.

We also find, as already discussed for year 2050 in Section 3.3.4, that the regional patterns of climate change in 2100, due to changes in emissions of short-lived gases and particles, are the result of regional patterns in the climate system's response rather than regional patterns in radiative forcing. The global patterns of surface temperature change in 2100 are similar for the short-lived gases and particles and the well-mixed greenhouse gases with the strongest surface temperature warming occurring over the summer continental United States and Mediterranean and the winter Arctic, while the major change in radiative forcing is over Asia (Levy et al., 2008). The predicted summertime warming over the United States is greatly enhanced by projected reductions in SO_2 emissions and increased black carbon emissions and the resulting positive radiative forcings over Asia. In the A1B scenario, this is assumed to be the result of Asian decisions addressing their local and regional air quality. The integrated assessment model projections for A1B assume that SO_2 emissions will be reduced in the future in order to improve air quality, but did

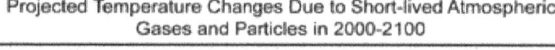
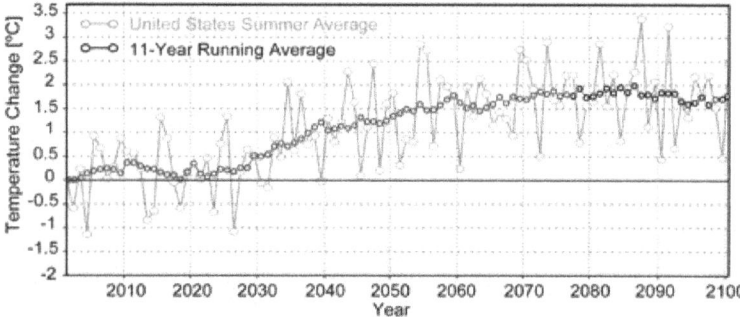

Figure 3.11 Surface temperature change in °C from 2000 over the twenty first century due to short-lived gases and particles over the continental United States during Northern Hemisphere summer in the GFDL model.

not explicitly project carbonaceous particle emissions. Scaling future carbonaceous emissions according to carbon monoxide emissions projections does not lead to similar reductions in emissions of these particles, so that there is an issue of consistency in projecting the influence of future air quality decisions that deserves further study.

3.4 REGIONAL EMISSIONS SECTOR PERTURBATIONS AND REGIONAL MODELS

3.4.1 Introduction to Regional Emissions Sector Studies

An additional set of simulations used global models to examine the impact of individual emissions sectors in specific regions on short-lived gases and particles. This study, in which the GISS and NCAR groups participated, was designed to examine the climate effects of short-lived gases and particles in a more policy-relevant way by focusing on the economic activities that could potentially be subject to regulation or reduction in usage (*e.g.*, by improved efficiency). We look at reductions in total emissions from a given sector in particular regions (North America and Asia), and do not consider any changes in technology or the relative contributions within a sector. As such, these are more useful for assessing the potential impacts of reductions in total power/fuel usage rather than changes in the mix of power generation/transportation types or in emissions control technologies targeted at specific pollutants.

3.4.2 Global Models

The GISS model setup for the regional emissions sector perturbation experiments was the same as that used in the transient climate studies (Section 3.2 and 3.3; see Appendix C and D, sections on Goddard Institute for Space Studies). The NCAR regional/sector perturbation simulations used the CAM-chem model (Lamarque *et al.*, 2005b), in which an updated version of the MOZART chemical transport model (Horowitz *et al.*, 2003) is embedded within the Community Atmosphere Model (CAM3, Collins *et al.*, 2006).

CAM-chem has a representation of tropospheric chemistry with non-methane hydrocarbons

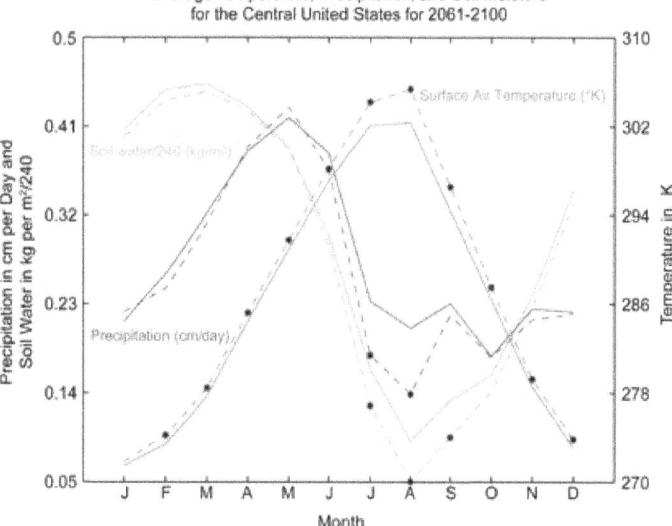

Figure 3.12 Monthly-mean time-series of available root-zone soil water (green lines, scaled by a factor of 1/240 for plotting purposes), precipitation (blue lines), and two-meter air temperature (red lines), averaged over the Central United States (105 to 82.5°W longitude; 32.5 to 45°N latitude). Dashed lines are for the ensemble mean of the A1B experiments, averaged over the years 2061 to 2100; solid lines are for the ensemble mean of the A1B* experiments, also averaged over the years 2061 to 2100. The asterisks represent those A1B monthly average values that are different from their companion A1B* values at the 95 percent confidence level. Temperature in K equals temperature in °C plus 273.15.

(NMHCs) treated up to isoprene, toluene and monoterpenes. The particle simulation in CAM-chem includes the bulk particle mass of black carbon (BC, hydrophobic and hydrophilic), primary organic carbon (POA, hydrophobic and hydrophilic), secondary organic carbon (SOA), ammonium and ammonium nitrate, and sulfate particles. Further details on the CAM-chem model are found in Appendix C in the section on National Center for Atmospheric Research.

3.4.3 Impact of Emissions Sectors on Short-Lived Gases and Particles

This set of experiments consisted of six simulations each reducing the present-day emissions by 30 percent in one sector for one region. By using present-day emissions, the results are not tied to any particular scenario. For present-day emissions, the IIASA 2000 inventory, based on the 1995 EDGAR 3.2 inventory extrapolated to 2000 using national and sector economic development data, was used (Dentener *et al.*, 2005), as in the GISS simulations described above. The exception to this is biomass burning emissions, which are taken from the Global Fire Emission Database (GFED) averaged over 1997

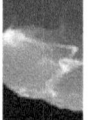

to 2002 (Van der Werf *et al.*, 2003) with emission factors from Andreae and Merlet (2001) for particles. The regions were defined as North America (60 to 130°W, 25 to 60°N) and Asia (60 to 130°E, 0 to 50°N) and the economic sectors defined according to the IIASA inventory as the domestic sector, the surface transportation sector, and a combined industry and power sector (the CAM-chem model did not perform the transportation sector simulations).

A control run with no perturbations was also performed to allow comparison. The goal of these simulations was to calculate the radiative forcing from all the short-lived gases and particles to identify the relative contribution of the given economic sectors in these two regions. This complements prior work examining the response to a subset of the gases and particles included here (*e.g.*, Koch *et al.*, 2007; Unger *et*

al., 2008). As the forcings were expected to be small, we concentrate on simple metrics rather than the climate response. The CAM-chem model did not calculate radiative forcing, so we also use aerosol optical depth, which is a good indicator of the radiative forcing from particles. All simulations were 11-year runs, with analysis performed over the last ten years.

The simulations were not performed using a full methane cycle, but the methane response to the imposed perturbations can be estimated by examining the changes in methane's oxidation rate. In these simulations, methane was prescribed at present-day values. Thus any change in methane oxidation is due solely to changes in the abundance of oxidizing agents. The difference in the steady-state abundance of methane that would occur as a result of this oxidation change is a simple calculation ($[CH_4]'/[CH_4] = L/L'$ for the global mean where L is the methane loss rate and the "prime" notation indicates the adjusted amounts). Use of the model's oxidation rate in the perturbation runs fully captures spatial and seasonal variations, and thus provides an accurate estimate of the equilibrium response of methane to the emissions changes. Finally, the radiative forcing resulting from these indirect methane changes is calculated using the standard formulation (Ramaswamy *et al.*, 2001).

We first examine the global mean annual average radiative forcing in the GISS model from the regional perturbations and those by economic sectors (Table 3.10). The effect of the

Table 3.10 Radiative forcing in milliWatts per square meter (mW per m²), from regional emission sector perturbations in the GISS model.

Region	Sector	Sulfate	BC	OC	Nitrate	Ozone	Methane (indirect)	All
North America	Domestic	0	-3	2	1	2	1	4
	Surface Transportation	-3	-5	0	1	-5	4	-9
	Industry/power	14	-2	-1	0	5	2	18
Asia	Domestic	0	-42	13	1	-12	-2	-41
	Surface Transportation	2	-8	1	2	-5	7	-2
	Industry/power	13	-4	0	-1	-1	5	12

Perturbations are 30 percent reduction in emissions of all species from the indicated economic sector in the given region. Direct forcings are shown for sulfate, black carbon (BC), organic carbon (OC), nitrate, and ozone. The effect of ozone precursor species on methane is included as methane "indirect". Note that particle indirect effects are not included.

Short-Lived Gas and Particle Annual Average Radiative Forcing (mW per m²)
due to 30 Percent Reduction in Emissions

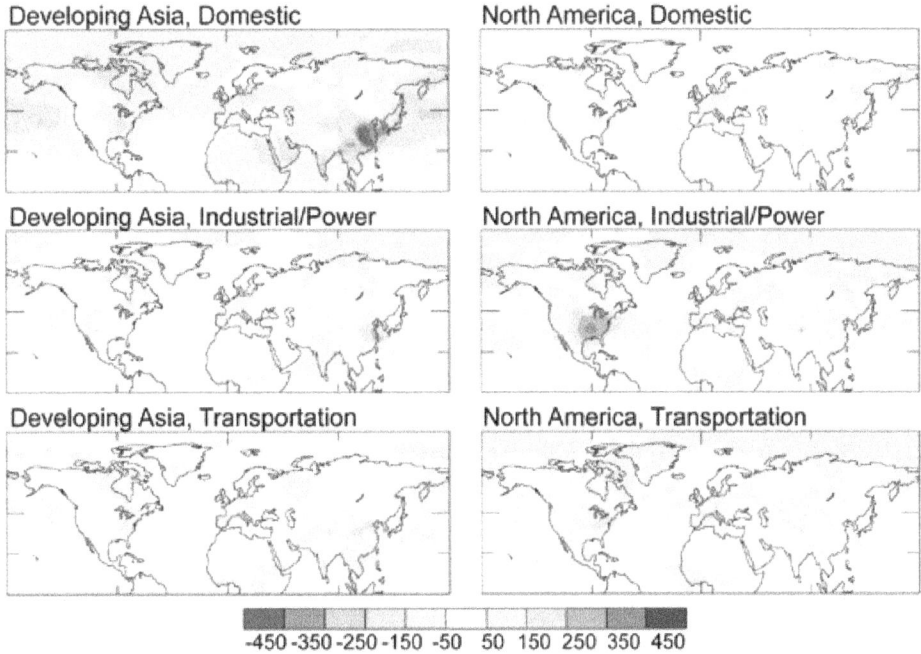

Figure 3.13 Short-lived gas and particle annual average radiative forcing (mW per m²) due to 30 percent reductions in emissions from the given region and economic sector in the GISS model.

perturbations is generally larger for Asian than for North American emissions. The only radiative forcing from an individual gas or particle to exceed 10 milliWatts per square meter (mW per m²) from a North American perturbation is the sulfate forcing from a reduction in industrial/power emissions. In contrast, forcings from sulfate, black carbon, organic carbon and ozone all exceed 10 mW per m² in response to perturbations in developing Asia, with the largest response for reductions in black carbon when domestic emissions are reduced (-42 mW per m²). The spatial pattern of the radiative forcing is also shown (Figure 3.13).

The two largest net forcings are in response to changes in North American industrial/power emissions, whose forcing is positive and is dominated by reductions in forcing from sulfate, and the Asia domestic sector, whose forcing is negative and dominated by reductions in black carbon and ozone. The spatial pattern of the aerosol optical depth changes capture the bulk of the radiative forcing in these two cases (Figure 3.14 *vs*. Figure 3.13). The sign is opposite, however, in the case of industrial/power emissions as these are dominated by reflective

sulfate particles, so decreased aerosol optical depth causes positive radiative forcing.

The GISS and CAM-chem models show very similar patterns of aerosol optical depth changes for these two perturbation experiments. For emissions reduction in the Asia domestic sector, the global mean aerosol optical depth decreases by 0.15 in the GISS model and 0.13 in the CAM-chem model, while for the North American industrial/power sector the decreases are 0.09 and 0.13, respectively. Hence the particle response appears to be fairly robust across these two models. Results suggest that the calculation of radiative forcing from aerosol optical depth introduces an additional inter-model difference that is less than that from the aerosol optical depth calculation (Schulz *et al.*, 2006), so that the total inter-model variation in radiative forcing from particles is probably on order of 50 percent. Results for ozone show marked differences, however, with the response of the tropospheric ozone column in the GISS model nearly always a factor of two to three greater than in the CAM-chem model. We believe that these differences primarily reflect the

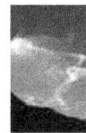

The effect of the perturbations is generally larger for Asian than for North American emissions.

Annual Average Aerosol Optical Depth Change Due to 30 Percent Reductions in Emissions

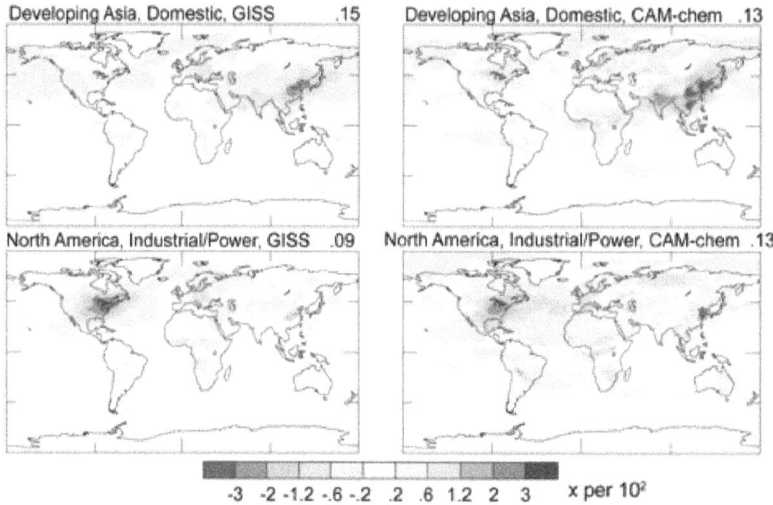

Figure 3.14 Annual average aerosol optical depth change due to 30 percent reductions in emissions from the given region and economic sector in the GISS (left column) and CAM-chem (right column) models. Values in the upper right give the global mean.

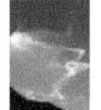

In most cases, substantial reduction in surface particle concentrations result from the regional economic sector emissions reductions. This is especially so in the Asia domestic analysis.

inclusion of the stratosphere in the GISS model, which leads to enhanced forcing as ozone near the tropopause has a particularly large radiative impact. Hence, the ozone radiative forcing is not yet robust to inter-model differences. However, particles typically have a larger influence on climate than ozone, so that the net radiative forcing remains a relatively more robust quantity.

We also examine changes in surface pollution levels in these simulations. Changes in surface ozone are typically small, with annual average local reductions of up to about 1 to 1.5 parts per billion by volume (ppbv) in both the CAM-chem and GISS models in response to reduction in transportation or industrial/power emissions. These increase to levels of 1 to 3 ppbv during boreal summer. Both these annual and summer increases are statistically significant. Changes in particles are larger. In most cases, substantial reduction in surface particle concentrations result from the regional economic sector emissions reductions. This is especially so in the Asia domestic analysis, where summer sulfate concentrations are lowered by 100 to 250 parts per trillion by volume (pptv) locally, and black carbon concentrations drop by 1800 to 2000 pptv for both summer and annual averages. Smaller air quality improvements are also clear

in the response to industrial/power and transportation emissions reductions in both regions. These reductions in particles are generally quite similar in the two models, with differences of only 5 to 20 percent in most cases.

The analysis shows that reductions in surface transportation emissions have a net negative forcing from short-lived gases and particles in both regions, primarily due to reductions in ozone and black carbon. As these are both pollutants at the surface, reducing emissions transport offers a way to simultaneously improve human health and mitigate climate warming (though the climate impact is quite small for Asia). The total climate mitigation would of course be larger adding in the effect of reduced emissions of long-lived greenhouse gases. In contrast, industrial/power sector emissions have their largest effect on climate through sulfate, and hence yield a positive forcing. Thus, the net effect of changes in short-lived gases and particles to industrial/power emissions reductions will offset a portion of any climate benefit from reduced emissions of long-lived gases. The domestic sector presents a similar picture to that seen for surface transportation. The effects are substantially larger in Asia, however. Hence, reductions in domestic emissions from Asia

Table 3.11 Total short-lived gas and particle radiative forcing (in mW per m²) as in Table 3.10 but for summer (June through August).

Region	Sector	Total forcing
North America	Domestic	6
	Surface Transportation	-10
	Industry/power	34
Asia	Domestic	-69
	Surface Transportation	-3
	Industry/power	10

Perturbations are 30 percent reduction in emissions of all species from the indicated economic sector in the given region. Direct forcings are shown for sulfate, black carbon (BC), organic carbon (OC), nitrate, and ozone. The effect of ozone precursor species on methane is included as methane "indirect". Note that particle indirect effects are not included.

offer another means to improve human health and mitigate warming. Note that the effects become particularly strong in Northern Hemisphere summer (Table 3.11), offering a potential path to mitigate increased summer heat over the Northern Hemisphere continents.

Overall, the Asia domestic emissions offer the strongest leverage on climate via short-lived gases and particles. This is partially a result of their magnitude, and partially their occurrence at lower latitudes than North American (or European) emissions. This enhances their impact as photochemistry is faster and incoming radiation is more abundant at lower latitudes. Perturbing the Asia domestic sector in the IIASA 2000 emissions inventory used here yields a much greater effect via black carbon changes than via sulfate changes. This reflects the influence of domestic fuel usage, for which black carbon is the dominant emission, and hence reductions from emissions in this sector in particular seem attractive for warming mitigation. As domestic usage and emissions are extremely difficult to quantify in the developing world, further studies of this sector are especially needed to characterize the uncertainty in these emissions. The GISS and CAM-chem results differ in the magnitude of the aerosol optical depth change resulting from the Asia domestic sector perturbations by only 13 percent, and this sector/ region has the largest influence in both models for both radiative forcing and surface pollution. The stronger aerosol optical depth response in

the CAM-chem model suggests that the radiative forcing in that model might be even larger than the 50 mW per m² global mean annual average seen in the GISS results.

Further work is required to more thoroughly characterize the robustness of these conclusions across a larger number of models, to explore the impact of particle indirect effects on clouds, and to examine alternative emissions scenarios considering changes in the mix of sources constituting a given sector and the influence of potential technological changes. The latter could be designed to reduce emissions of particular pollutants, while not affecting others. Our results for the radiative forcing from individual gases and particles give an idea of the potential impact of such technologies. However, we note that these technologies could also have effects on overall fuel consumption by altering the efficiency of a particular process.

Interestingly, both the transient climate projections and the present-day perturbations find that emissions from Asia are the most important controllers of climate trends or mitigation. Given that the radiative forcing reduction from decreases in Asia domestic emissions extends over much of the Northern Hemisphere (Figure 3.13), and the conclusion from the transient

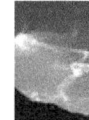

these levels vary spatially and are affected by local emissions, regional scale models are needed to develop the emissions control strategies that meet these standards at county and state levels. To achieve reductions in small-diameter particles, emissions scenarios often include utility sector emissions reductions to lower sulfate particle levels. As shown in the previous section, lowering sulfate particle concentrations could actually have negative implications for radiative forcing and climate temperature increases. Regional downscaling studies shown in this section suggest that future changes in regional climate could reduce the benefits from anticipated emissions reductions on lowering ozone.

Downscaled regional scale climate simulations (*e.g.*, Leung and Gustafson, 2005; Liang *et al.*, 2006; Liang *et al.*, 2004) rely on a global climate model to provide boundary conditions for the regional domain as well as the radiative effect of well-mixed greenhouse gases within the domain for the radiation calculations. Regionally downscaled climate simulations are needed by a number of applications that must consider local changes in future climate. Since ozone and small particles exceedances of regulatory thresholds are substantially affected by local scale changes in emissions and meteorology, several recent studies using regionally downscaled climate scenarios have been used to study the sensitivity of air quality to potential changes in future climate. The primary purpose of these studies was to study how increases in temperature and other future climate changes could affect ozone and small-diameter particles and potentially decrease the effectiveness of anticipated emissions reductions.

climate simulations that the climate response to short-lived gas and particle changes is not closely localized near their emissions, it seems plausible that emissions from this region may have as large or larger an effect on other parts of the Northern Hemisphere as changes in local emissions.

3.4.4 Regional Downscaling Climate Simulations

The sector-based simulations presented in Section 3.4.3 suggest that reductions in surface level ozone or black carbon would have a negative radiative forcing, while reductions in sulfate particles from the utility sector would have a positive radiative forcing. If concentration levels for ozone and small-diameter particles ($PM_{2.5}$)[6] exceed threshold standards under the United States Clean Air Act, emissions control strategies must be developed. Since

Downscaled regional climate model simulations by Gustafson and Leung (2007) and Nolte *et al.* (2008) have used the EPA/NOAA air pollution model to test the impact of future (*ca.* 2050) climate on ozone and particles with current emissions scenarios. Biogenic emissions, which are meteorologically dependent, were recalculated

---

[6] $PM_{2.5}$ are all small particles with diameters less than 2.5 micrometers

Change in 95th Percentile Ozone
June • July • August

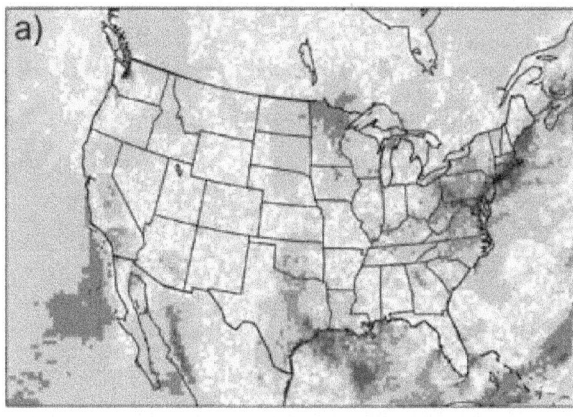

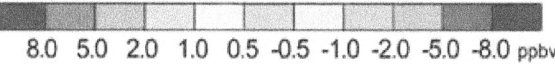

8.0 5.0 2.0 1.0 0.5 -0.5 -1.0 -2.0 -5.0 -8.0 ppbv

Change in 95th Percentile Ozone
September • October

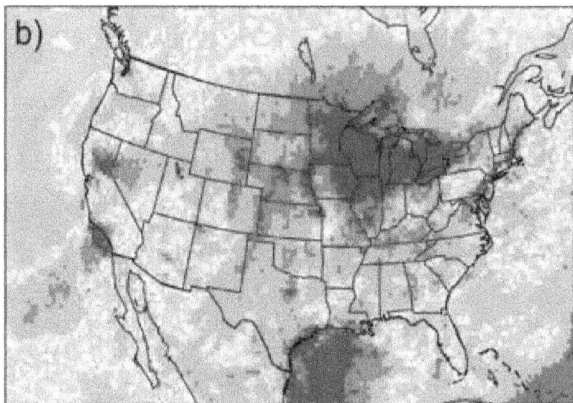

Figure 3.15 From Nolte *et al.* (2008), the change in ozone at the upper end of the ozone distribution (average of the ≥95th percentile values for each grid) for (2050 to present) years of simulation under AIB regional climate model simulations.

be considered in air quality management.

These regional downscaling climate studies rely on climate forcing linkages from global climate simulations with future trends for long-lived gases including CO_2, CH_4, N_2O, and halocarbons. The influence of short-lived gases and particles on future climate has not been included in those studies to date; however, more recent developments are underway to include direct and indirect radiative effects in the regional chemistry model. Based on the known positive radiative forcing effect of ozone, the increases in ozone in response to future climate in Figure 3.15 should have a positive radiative forcing that could dampen the net negative radiative forcing anticipated from future emissions reductions for ozone (Section 3.4.3).

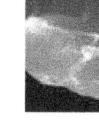

While regional downscaling climate impacts from short-lived gases and particles cannot be directly reported on, future emissions scenarios were considered by several regional downscaling studies for the purpose of air quality impacts under future climate. The impact of climate only and then an emissions change scenario were tested by Nolte *et al.* (2008) and Hogrefe *et al.* (2004). As presented in Figure 3.15, results looking only at the climate change without future changes in emissions suggest that future climate changes could increase maximum ozone levels by approximately eight ppbv or ten percent in some regions of North America. Looking at future emissions scenarios for nitrogen oxides and sulfur dioxide reductions demonstrates that the uncertainty in the future emissions scenarios

under the future climate scenario. Results suggest that future climate changes could increase maximum ozone levels by approximately ten percent in some regions (Figure 3.15). With anticipated emissions reductions under United States Clean Air Act requirements, these results suggest that future climate could dampen the effectiveness of these emissions controls. Evaluation of ensemble regional climate model results are essential for this application before quantitative conclusions can be made about the impact of future climate on specific emissions control strategies; however, these results suggest that climate change is a factor that needs to

Climate change itself could have a significant impact on low-level ozone concentrations, leading to yet further changes in climate.

introduces a much larger variation in the air quality conclusions depending on the scenario (Hogrefe *et al.*, 2004; Nolte *et al.*, 2008). Similar to the findings here about short-lived gases' and particles' impact on climate, the range of plausible air quality impacts from future emissions scenarios suggest very different outcomes, and the future scenarios of emissions for short-lived gases and particles have a great deal of obvious, inherent uncertainty.

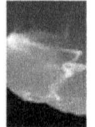

CHAPTER 4

Findings, Issues, Opportunities, and Recommendations

Lead Authors: Hiram Levy II, NOAA/GFDL; Drew T. Shindell,
NASA/GISS; Alice Gilliland, NOAA/ARL

4.1 INTRODUCTION

This Chapter, which is intended for both technical and non-technical audiences, provides a summary of the key findings, presents a number of new questions that were revealed by our study, and identifies new opportunities for future research.

4.2 KEY FINDINGS

The key findings of Synthesis and Assessment Product 3.2 are summarized below:

1. Our results suggest that the projected changes in short-lived[1] radiatively active[2] gases and particles may significantly influence the climate, even out to year 2100. Their projected changes can account for up to 40 percent of the projected warming over the summertime continental United States by 2100.

2. We find that spatial patterns of radiative forcing[3] and climate response are quite distinct. Thus, both short-lived and long-lived[4] gases and particles cause enhanced climate responses in the same regions rather than short-lived gases and particles having an enhanced effect primarily in or near polluted areas. This means that regional pollution control strategies will have large-scale climate impacts.

3. Reductions of short-lived gas and particle emissions from the domestic fuel-burning sector in Asia appear to offer the greatest potential for substantial, simultaneous

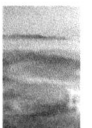

---

[1] Atmospheric lifetimes for the short-lived radiatively active gases and particles of interest in the lower atmosphere are about a day for nitrogen oxides, a day to a week for most particles, and a week to a month for ozone. As a result of their short lifetimes, their concentrations are highly variable in space and time and they are concentrated in the lowest part of the atmosphere, primarily near their sources.

[2] "Radiatively active" indicates the ability of a substance to absorb and re-emit radiation, thus changing the temperature of the lower atmosphere.

[3] Radiative forcing is a measure of how the energy balance of the Earth-atmosphere system is influenced when factors that affect climate, such as atmospheric composition or surface reflectivity, are altered. When radiative forcing is positive, the energy of the Earth-atmosphere system will ultimately increase, leading to a warming of the system. In contrast, for a negative radiative forcing, the energy will ultimately decrease, leading to a cooling of the system. For technical details, see Box 3.2.

[4] Atmospheric lifetimes for the long-lived radiatively active gases of interest range from ten years for methane to more than 100 years for nitrous oxide. While carbon dioxide's lifetime is more complex, we can think of it as being more than 100 years in the climate system. As a result of their long atmospheric lifetimes, they are well-mixed and evenly distributed throughout the lower atmosphere. Global atmospheric lifetime is the mass of a gas or a particle in the atmosphere divided by the mass that is removed from the atmosphere each year.

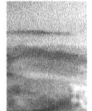

improvement in local air quality and reduction of global warming. Reduction in emissions from surface transportation would have a similar impact in North America.

4. The three comprehensive climate models[5], their associated chemical composition models[6] and the different projections of short-lived emissions all lead to a wide range of projected climate impacts by short-lived gases and particles. Each of the three studies represents a thoughtful, but incomplete, characterization of the driving forces and processes that are believed to be important to the climate or to the global distributions of the short-lived gases and particles. Much work remains to be done to characterize the sources of the differences and their range.

5. The two most important uncertainties in characterizing the potential climate impact of short-lived gases and particles are found to be the projection of their future emissions and the determination of the indirect effect of particles[7] on climate. See Section 4.3 for a discussion of the fundamental difference between uncertainties in future emissions and uncertainties in processes such as the indirect effect of particles on climate.

6. Natural particles such as dust and sea salt also play an important role and their emissions and interactions differ significantly among the models, with consequences to the role of short-lived pollutants. This

---

[5] Comprehensive climate models are state-of-the-art numerical representations of the climate based on the physical, chemical, and biological properties of its components, their interactions and feedback processes that account for many of the climate's known properties. Coupled Atmosphere-Ocean (-sea ice) General Circulation Models (AOGCMs) provide a comprehensive representation of the physical climate system.

[6] Chemical composition models are state-of-the-art numerical models that use the emission of gases and particles as inputs and simulate their chemical interactions, global transport by the winds, and removal by rain, snow, and deposition to the earth's surface. The resulting outputs are global three-dimensional distributions of the initial gases and particles and their products.

[7] Particles may have an indirect effect on the climate system by modifying the optical properties and lifetime of clouds. A detailed technical discussion is given in Box 3.1.

inconsistency among models should be addressed in future studies.

7. The SAP 2.1a emissions scenarios (Clarke *et al.*, 2007) for long-lived gases produce climate projections that are within the IPCC range, although it should be noted that the lower-bound stabilization emissions scenario[8], which is equivalent to a carbon dioxide stabilization level of 450 parts per million, results in global surface temperatures below those calculated for the IPCC storyline emissions scenario with the most moderate carbon dioxide increases (B1), particularly beyond 2050.

4.3 ISSUES RAISED

It is important to recognize the difference between uncertainties in processes and uncertainties in future emissions. Uncertainties in chemical and physical processes, which are discussed in Sections 4.3.2 and 4.3.3, represent the state of our current knowledge. The fact that one modeling group chooses to include a process such as indirect forcing of climate by particles, while another group chooses not to, shows that our knowledge about short-lived gases and particles and their interactions with climate is still evolving. Eventually, with further research, uncertainties in chemical and physical processes can be significantly reduced. However, uncertainties in future emissions, which are discussed in Section 4.3.1, will always be with us. What we can do is develop a set of internally consistent emissions scenarios that include all of the important radiative gases and particles and bracket the full range of possible future outcomes.

4.3.1 Emissions Projections

The analysis presented in Chapter 3 showed that the main contributors to the divergence among model projections of future particle loading and climate forcing were the differences in the underlying emissions projections.

---

[8] Stabilization emissions scenarios are a representation of the future emissions of a substance based on a coherent and internally consistent set of assumptions about the driving forces (such as population, socio-economic development, technological change) and their key relationships. These emissions are constrained so that the resulting atmospheric concentrations of the substance level off at a predetermined value in the future.

Those differences arose from two primary factors. First, different integrated assessment models[9] interpret a common socio-economic 'storyline' in different ways, as demonstrated in Chapter 2. Second, emissions scenarios were not produced for some short-lived gases and particles and had to be added in later by other emissions modeling groups or by the climate modeling groups themselves. These same issues were also encountered in Chapter 2 which focused on the SAP 2.1a stabilization emissions scenarios (Clarke *et al.*, 2007). While emissions scenarios for short-lived gases and particles were outside of SAP 2.1a's mandate, climate projections require them. Part of the reason for the different emission inventories used here and in the IPCC studies was that the integrated assessment models did not recognize that they were necessarily important when the scenarios were first constructed.

Just consider two of the integrated assessment models used to generate the sulfur dioxide and nitrogen oxide emissions for the A1B scenario used in Chapter 3. The two models project different rates of growth; total energy use is also different in the two models, with 3 percent greater use in one model by 2030, but 9 percent less usage at 2050. That same model is less optimistic about emissions controls.

None of the emissions models predict black and organic carbon particle emissions. The GFDL composition model followed the IPCC suggestion to scale future carbon particle emissions by the emissions for carbon monoxide, which are projected to increase throughout the twenty-first century. By 2050, the projected carbon monoxide emissions increase ranges from 8 percent to 119 percent across the collection of integrated assessment models used for the last IPCC report. In contrast, the GISS global chemical composition model used recent estimates by Streets *et al.* (2004) that project a substantial decrease in future emissions of carbon particle (Figure 3.1; Table 3.1). Ammonia emissions present a similar problem. They are sometimes scaled by default to follow nitrous oxide, which

is projected to increase significantly. Given the number of ammonia sources that are disconnected from nitrous oxide production, this may be questionable. Moreover, newer projections for ammonia emissions have a much slower rate of increase.

Finally, the global chemical composition models all used their own natural emissions. Though these were held constant, they influence the response to anthropogenic emissions by determining the background abundance of short-lived gases and particles. The level of a natural particle, dust, can also directly affect the level of an anthropogenic particle, as it does in the GISS model by removing sulfate.

We face very significant problems in projecting the future emissions for short-lived gases and particles. Future climates are only weakly dependent on projected emissions for the next 20 years when, due to the inertia in major emitters, we may have credibility in forecasting emissions trends for short-lived gases and particles such as the nitrogen oxides and sulfur dioxide. However, we have shown that plausible emissions scenarios have the potential for significant impacts on climate through the rest of the twenty-first century.

Unlike the long-lived greenhouse gases, short-lived gases and particles do not accumulate, so their full impact at year 2100 depends on end of the century emissions. At this time, there is no credible quantitative skill in forecasting these emissions out to 2100. As Chapter 3 demonstrates, it is not even clear that we can currently predict the sign of the change for black carbon emissions over the next decade. This is a problem that requires not just enhanced scientific knowledge, but also the ability to predict social, economic and technological developments as far as 100 years into the future. One needs only to think back to 1907 to realize how difficult that is and will be.

4.3.2 Particles (Indirect Effect, Direct Effect, Mixing, Water Uptake)

We find that several aspects of particle modeling have large uncertainties (see Figure SPM.2 in IPCC Fourth Assessment Report [IPCC, 2007] for a qualitative estimate of these uncertainties), of which the particle indirect effect, which is

---

[9] Integrated assessment models are a framework of models, currently quite simplified, from the physical, biological, economic, and social sciences that interact among themselves in a consistent manner and can be used to evaluate the status and the consequences of environmental change and the policy responses to it.

very poorly known, is probably the most critical. Many aspects of the particle-cloud interaction are not well quantified, and hence the effect was left out entirely in the GFDL and CCSM simulations. The GISS model used a highly parameterized approach that is quite crude. The modeling community as a whole cannot yet produce a credible characterization of the climate response to particle/cloud interactions. Moreover, the measurements needed to guide this characterization do not yet exist. All mainstream climate models (including those participating in this study) are currently either ignoring it, or strongly constraining the model response. Attempts have been made using satellite and ground-based observations to improve the characterization of the indirect effect, but major limitations remain.

As discussed in Section 3.2.4, observations of aerosol optical depth[10] are best able to constrain the total extinction (absorption plus scattering) of sunlight by all particles under clear-sky conditions, but not to identify the effect of individual particles which may scatter (cool) or absorb (warm). Improved measurements of extinction and absorption may allow those two classes of particles to be separated, but will not solve the fundamental problem of determining their relative individual importance. As seen in this and other studies, models exhibit a wide range of relative contributions to total aerosol optical depth from the various natural and anthropogenic particles (Figure 3.2). Thus, the direct radiative effect of changes in a particular particle can be substantially different among models depending upon the relative importance of that particle.

---

[10] Aerosol optical depth is a measure of the fraction of the sun's radiation at a given wavelength absorbed or scattered by particles while that radiation passes through the atmosphere.

Additionally, particles are not independent of one another. They mix together, a process that is only beginning to be incorporated in composition and climate models. In these studies, for example, the GISS model included the influence of sulfate particles sticking to dust, which can decrease the sulfate radiative forcing, by ~40 percent between 2000 and 2030 (Bauer *et al.*, 2007), but the sticking rates are quite uncertain. Mixing of other particle types is also highly uncertain, but is known to occur in the atmosphere and would also affect the magnitude of particle radiative forcings.

Another process that influences the effect of particles on climate is their uptake of water vapor, which alters their size and optical properties. This process is now included in all state-of-the-art comprehensive climate models. As the uptake varies exponentially with relative humidity, small differences in treatment of this process have the potential to cause large discrepancies. However, our analysis in Chapter 3 (*e.g.,* Table 3.7) suggests that the differences induced by this process may be small relative to the others we have just discussed.

4.3.3 Climate and Air Quality Policy Interdependence

Chapter 3 exposes major uncertainties in the climate impacts of short-lived gases and particles that will have to be addressed in future research. We raise important issues linking air quality control and global warming, but are unable to provide conclusive answers. We are able, however, to identify key questions that must be addressed by future research.

Most future sources of short-lived gases and particles result from the same combustion processes responsible for the increases in atmospheric carbon dioxide. However, while reductions in their emissions are currently driven by local and regional air pollution issues that can be addressed independently of any reductions in carbon dioxide emissions, in the future a unified approach could effectively address both climate and air quality issues. Furthermore, the climate responses to emissions changes in short-lived pollutants can be felt much more quickly because of shorter atmospheric lifetimes. The good news is that there is at least one clear win-win solution for climate (less warming) and air quality (less pollution): methane reduction. Decreases in methane emissions lead to reduced levels of lower atmospheric ozone, thereby improving air quality; and both the direct methane and indirect ozone decreases lead to reduced global warming (Fiore *et al.*, 2002; Shindell *et al.*, 2005; West and Fiore, 2005; West *et al.*, 2006). Reductions in emissions of carbon monoxide or volatile organic compounds (VOCs) have similar effects, namely leading to reduced abundances of both methane and ozone (Berntsen *et al.*, 2005; Shindell *et al.*, 2005; West *et al.*, 2006; West *et al.*, 2007), therefore providing additional win-win strategies for improvement of climate and air quality. Reductions in black carbon particles and nitrogen oxide are potentially win-win as well, but the climate impact of reductions in their emissions is uncertain. On the other hand, the reduction of sulfur and organic carbon particles results in a reduction of cooling and increased global warming.

The cases of black carbon (soot) and nitrogen oxide gases are illustrative of the complexities of this issue. A major source of soot is the burning of biofuel, the sources of which are primarily animal and human waste as well as crop residue, all of which are considered carbon dioxide neutral (*i.e.,* the cycle of production and combustion does not lead to a net increase in atmospheric carbon dioxide). Current suggested replacements result in the release of fossil carbon dioxide. Therefore this reduction in biofuel burning, while reducing the emission of soot, will increase the net emission of carbon dioxide. The actual net climate response from reduced use of biofuel is not clear. The case of nitrogen oxides appears to be approximately neutral for climate, though clearly a strong win for air quality. Reducing nitrogen oxides reduces ozone, which reduces warming. However, reductions in both lead to reduced hydroxyl radicals and therefore an increased level of methane, which increases warming.

There clearly are win-win, win-uncertain, and win-lose situations regarding climate and actions taken to improve air quality. We are not making any policy recommendations in Synthesis and Assessment Product 3.2, but we do identify the policy relevant scientific issues. At this time we can not provide any quantita-

tively definitive scientific answers beyond the well known facts that the decrease of sulfur and organic carbon particles, both of which cool the climate, will increase global warming, while decreased methane, carbon monoxide, and volatile organics will decrease global mean warming. Decreases in the burning of biofuel, as well as decreased emissions of nitrogen oxides, are more complex and the net result is not clear at this time.

4.4 RESEARCH OPPORTUNITIES AND RECOMMENDATIONS

This last section of the report is a call for focused scientific research in emissions projections, radiative forcing, chemical composition modeling and regional downscaling. Particular emphasis needs to be paid to the future emissions scenarios for sulfur dioxide, black carbon particles and nitrogen oxides, to the indirect radiative forcing by particles, and to a number of ambiguities in current treatments of transport, deposition, and chemistry.

4.4.1 Emissions Scenario Development

Future climate studies must seriously address the very difficult issue of producing realistic and consistent 100-year emissions scenarios for short-lived gases and particles that include a wide range of socio-economic and development pathways and are driven by local and regional air quality actions taken around the globe.

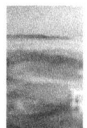

The current best projections used in this report and in the Fourth Assessment Report of the IPCC do not even agree on whether black carbon particle and nitrogen oxide emissions trends continue to increase or decrease. While all the current sulfur dioxide emissions projections used in this study assume that emissions in 2100 will be less than at present, how much less is quite uncertain, and all of these projected decreases by 2100 may well be wrong. Part of the reason for the different emission inventories used here and in the IPCC studies was that the integrated assessment models did not recognize that these gases and particles were necessarily important when the scenarios were first constructed. Clarification of the challenges associated with emissions projections (not a simple matter of improving quantitative skill, as these are a function of difficult-to-anticipate socio-economic choices) is also necessary.

As the greatest divergences in our study came from the carbon particles that were not projected for the A1B scenario, we strongly recommend that future emissions scenarios pay greater attention to them and provide consistent emissions projections for carbon particles and ammonia along with the other gases and particles. We are aware that many integrated assessment models are already capable of providing this information (*e.g.,* two of the three discussed in Chapter 2 provide carbon particle emissions).

We also recommend that climate models make greater efforts to study the effects of short-lived emissions projections in a manner that isolates their effect from that of the long-lived greenhouse gases. In particular, we believe there is merit in continuing to use a broad distribution of integrated assessment models to realistically characterize the range of potential futures for a given socio-economic storyline. In order to understand the contribution to uncertainty by the composition and climate models, it would also be worthwhile to perform a controlled experiment with identical emissions projections using multiple composition and climate models.

4.4.2 Particle Studies (Direct Effect, Indirect Effect, Mixing, Water Uptake)

Calculation of the indirect effect is potentially the single most important deficiency in this study that can be directly improved. None of the models in the IPCC Fourth Assessment Report study or in this Product realistically accounted for the full particle indirect radiative forcing.

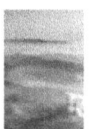

Given that the inclusion of a crude treatment of the particle indirect effect played a substantial role in one model's response in this study, it is clear that better characterization of this effect is imperative. The development of a measurement program to characterize particle indirect effects is a critical need.

It is also clear that other potential particle processes need to be examined. An example is interactive dust loading, which can influence the composition of other short-lived gases and particles and can also be influenced by them (*e.g.,* via changes in solubility due to acid uptake [Fan *et al.*, 2004; Bauer and Koch, 2005]). Dust emissions will also respond to vegetation changes as the climate warms. It has been speculated that arid regions may contract as a result of fertilization of plants by increased CO_2, reducing emissions (Mahowald and Luo, 2003), while source regions may expand as a result of global warming and reduced rainfall, and thus increase emissions (Woodward *et al.*, 2005). Note that the actual trend will depend upon local changes in climate, especially rainfall, which is among the least robust aspect of current climate projections. Other processes of potential importance that were not included in these transient climate simulations are changes in atmospheric levels of sea-salt particles and changes in darkening of snow and ice surfaces by soot deposition.

Additional observations are clearly needed to better constrain the optical properties of aerosols. Measurement by devices such as the aerosol polarimeter on the NASA satellite, Glory, should provide some of the needed information. We recommend emphasis on long-term particle monitoring from ground and space, and on better characterization of particle interactions with clouds in the laboratory. We also recommend greater use of the distinction between scattering and absorbing particles to characterize their relative importance.

4.4.3 Improvements in Transport, Deposition, and Chemistry

The emissions issues become even more problematic when the future distributions employed in the comprehensive climate models are generated by multiple global composition models, all with differing treatments of mixing in the lowest layers of the atmosphere, different treatment of transport and mixing by turbulence and clouds, different treatments of the removal of gases and particles by rain, snow and contact with the earth's surface, and different approximate treatments of the very large collection of chemical reactions that we do not yet fully understand. There are research opportunities in the study of all phases of the physical and chemical processes that determine the global distribution of short-lived gases and particles from their emission through their removal from the atmosphere.

4.4.4 Recommendations for Regional Downscaling

Regional downscaling, where global climate models introduce climate forcing into regional climate models, is a relatively new development. Current regional downscaling results have relied on older comprehensive climate model simulations. Data from the newer comprehensive global models are needed and coordination and planning are critical since downscaling require input data every three or six hours. While this will lead to huge amounts of data to be stored and transferred and poses a non-trivial technical problem, it is essentially an infrastructure/practical restriction rather than a source of scientific challenges. A carefully coordinated set of region climate model predictions using various global and regional scale models and future scenarios is needed to reduce the uncertainty and identify methodological improvements.

The North American Regional Climate Change Assessment Program (NARCCAP, <http://www.narccap.ucar.edu>) is an ongoing effort that is actively taking this approach of multiple regional climate model simulations and multiple future scenarios. Four different comprehensive global models and six different regional models are being used to intercompare and evaluate regional climate simulations for North America. This effort and others like it should greatly help to advance regional downscaling approaches and provide important model archives for environmental quality and resources applications.

It is important to also note that these studies do not currently include short-lived gases and particles in the global climate simulations. Further, many regional models do not include feedbacks between air quality, the radiation budget and climate. These feedbacks may be quite important. For example, The U.S. EPA Clean Air Interstate Rule requires almost a 30 percent reduction in sulfur dioxide emissions in the eastern United States during the next two decades. This could have a significant impact on the climate projections, as was shown in Chapter 3. We also need to separate the impacts on regional climate by the direct and indirect effects of particles in general from the impacts on regional climate by local emissions changes.

More research is clearly needed to determine if downscaled regional climate simulations actually provide more detailed and realistic results. The higher regional resolution is important for a wide range of environmental issues including water and air quality, agricultural productivity, and fresh water supplies. For example, highly resolved regional climate information is needed to accurately predict levels of ozone and small particles, which are strongly influenced by local changes in emissions and climate.

4.4.5 Expanded Analysis and Sensitivity Studies

Analyses of surface temperature response to changes in short-lived gases and particles need to be strengthened by additional sensitivity studies that should help to clarify causes and mechanisms. For example, in the GISS model, how much warming did the declining trend in the indirect effect contribute to its climate response and where? How would the GISS results differ if dust had not been permitted to take up sulfate particles? Determining the relative importance of these and other processes to the climate response would help prioritize the gaps in our knowledge. There are also a wide range of climate-chemistry feedbacks and controls that should be explored. Both the response of the climate system to controls on short-lived gases and particles and the possible feedbacks, and the possible impacts of climate changes on levels of short-lived gases and particles are all fertile areas for future research. While it was not possible, both due to time and resource restraints, for this study to explore these additional analyses, we recommend their future study.

The major unfinished analysis question in this study is the relative contribution of a model's

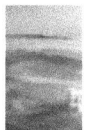

regional climate response, as opposed to the contribution from the regional pattern of radiative forcing, to the observed regional change in seasonal surface temperature. (See Section 9.2.2.1 of Chapter 9 in the IPCC Fourth Assessment Report [Hegerl *et al.*, 2007] and references therein for further discussion of this issue.) Specific modeling studies are needed to answer questions such as: Is there a model independent regional climate response? and What are the actual physical mechanisms driving the regional surface temperature patterns that we observe? These questions appear to identify a very important area of study, particularly given the apparently strong climate response in the summertime central United States.

4.5 CONCLUSION

After considering all of the issues discussed in this chapter, the net result is that we agree that short-lived species may have significant impact on future climate out to 2100. However, we could not find a consensus in this report on the duration, magnitude or even sign (warming or cooling) of the climate change due to future levels of the short-lived gases and particles. We have presented a plausible case for enhanced climate warming due to air quality policies that focus primarily on sulfate particle reduction and permit the emission of soot to continue to increase as realized in a version of the IPCC's A1B scenario. Alternative versions of this scenario that follow different pollution control storylines could have less impact. While we do not have definitive answers to the second goal of this report to "assess the sign, magnitude, and duration of future climate changes due to changing levels of short-lived gases and particles that are radiatively active and that may be subject to future mitigation actions to address air quality issues," we do provide plausible estimates that begin to characterize the range of possibilities and we identify key areas of uncertainty and provide motivation for addressing them.

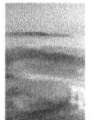

IPCC Fourth Assessment Report Climate Projections
(Supplemental to Chapter 2)

Lead Author: Hiram Levy II, NOAA/GFDL

Contributing Author: Tom Wigley, NCAR

These robust conclusions, which we believe also apply to the climate projections from the SAP 2.1a scenarios (Clarke *et al.*, 2007), are taken primarily from the Executive Summary of Chapter 10 of the IPCC's Fourth Assessment Report (Meehl *et al.*, 2007) as well as some details extracted from the body of Chapter 10, and are summarized below.

A.1 MEAN TEMPERATURE

All Atmosphere-Ocean General Circulation Models (AOGCMs) in Chapter 10 of the Fourth Assessment Report (AR4) (Meehl *et al.*, 2007) project increases in global mean surface air temperature (SAT) throughout the twenty-first century, with the warming proportional to the associated radiative forcing. There is close agreement among globally averaged SAT multi-model mean warming for the early twenty-first century for the three SRES (B1, A1B and A2) scenarios as well as for SAP 2.1a Level 2 through 4 scenarios out to 2050. The warming rate over the next few decades in Chapter 10 (Meehl *et al.*, 2007) is affected little by different scenario assumptions or different model sensitivities, and is similar to that observed for the past few decades. By mid-century (2046 to 2065), the choice of SRES scenario becomes more important and they start to separate, though the range among the collection of AOGCMs is comparable. By the end of the twenty-first century, the SATs generated by MAGICC using the 12 SAP 2.1a scenarios in Chapter 2 as well as the full spread of all of the AOGCMs for the A2, B1 and Committed projections have completely separated, though A1B still has some overlap with A2 and B1.

In general, geographical patterns of projected SAT warming show greatest temperature increases over land (roughly twice the global average temperature increase) and at high northern latitudes, and show less warming over the southern oceans and North Atlantic, consistent with observations during the latter part of the twentieth century. These patterns are similar across the B1, A1B, and A2 scenarios (see Figure 10.8 in Chapter 10 of the AR4; Meehl *et al.*, 2007) only increasing in magnitude with increasing radiative forcing. Results for the stabilization emission scenarios similar to those studied here should show the same pattern similarities at least out to 2100 (Dai *et al.*, 2001a,b). It should be noted that, in none of the cases considered here, has the climate stabilized by 2100; for the higher stabilization levels this may take centuries. Temperature change patterns may differ as one approaches closer to a stable climate.

A.2 TEMPERATURE EXTREMES

It is very likely that heat waves will be more intense, more frequent and longer lasting in a future warmer climate. Cold episodes are projected to decrease significantly in a future warmer climate. Almost everywhere, daily minimum temperatures are projected to increase faster than daily maximum temperatures, leading to a decrease in diurnal temperature range. Decreases in frost days are projected to occur almost everywhere in the mid and high latitudes, with a comparable increase in growing season length (Meehl *et al.*, 2007).

A.3 MEAN PRECIPITATION

Globally averaged mean atmospheric water vapor, evaporation and precipitation are projected to increase. By 2100, precipitation generally increases in the areas of regional tropical precipitation maxima (such as the monsoon regimes) and over the tropical Pacific in particular, with general decreases in the subtropics, and increases at high latitudes as a consequence of a general intensification of the global hydrological cycle. The geographical patterns of precipitation change during the twenty-first century are not as consistent across AOGCMs and across scenarios as they are for surface temperature (Meehl *et al.*, 2007).

A.4 PRECIPITATION EXTREMES AND DROUGHTS

Intensity of precipitation events is projected to increase, particularly in tropical and high latitude areas that experience increases in mean precipitation. There is a tendency for drying of the mid-continental areas during summer, indicating a greater risk of droughts in those regions. Precipitation extremes increase more than the mean in most tropical and mid- and high-latitude areas (Meehl *et al.*, 2007).

A.5 SNOW AND ICE

As the climate warms, snow cover and sea ice extent decrease; glaciers and ice caps lose mass owing to dominance of summer melting over winter precipitation increases. There is a projected reduction of sea ice in the twenty-first century both in the Arctic and Antarctic with a large range of model responses. Widespread increases in thaw depth over much of the permafrost regions are projected to occur in response to warming over the next century (Meehl *et al.*, 2007).

Note: All of the AR4 predictions for precipitation, snow cover, and sea and land ice are less certain and more variable across the suite of AOGCMs than they are for both the global average and the more robust geographic patterns of temperature.

A.6 CARBON CYCLE

Under the SRES illustrative emissions scenarios, for central carbon-cycle model parameters, CO_2 concentrations are projected to increase from its present value of about 380 ppm to 540 to 970 ppm by 2100. The SAP 2.1a Reference scenarios give 2100 concentrations of 740 to 850 ppm (Clarke *et al.*, 2007). There is unanimous agreement among the simplified climate-carbon cycle models that future climate change would reduce the efficiency of the Earth system (land and ocean) to absorb anthropogenic carbon dioxide. The higher the stabilization emission scenario warming, the larger is the impact on the carbon cycle. Both MAGICC and two of the three integrated assessment models used in SAP 2.1a contain simplified carbon cycle models comparable to those in Chapter 10 of the AR4 (Meehl *et al.*, 2007).

A.7 OCEAN ACIDIFICATION

Increasing atmospheric CO_2 concentrations lead directly to increasing acidification of the surface ocean. Multi-model projections based on SRES scenarios give reductions in pH of between 0.14 and 0.35 units over the twenty-first century, adding to the present decrease of 0.1 units from preindustrial times.

A.8 SEA LEVEL

"Sea level is projected to rise between the present (1980 to 1999) and the end of this century (2090 to 2099) under the SRES B1 scenario by 0.28 m for the multi-mode average (range 0.19 to 0.37 m), under A1B by 0.35 m (0.23 to 0.47 m), under A2 by 0.37 m (0.25 to 0.50 m) and under A1FI by 0.43 m (0.28 to 0.58 m). These are central estimates with 5 to 95 percent intervals based on AOGCM results, not including uncertainty in carbon-cycle feedbacks. In all scenarios, the average rate of rise during the twenty-first century very likely exceeds the 1961 to 2003 average rate (1.8 ± 0.5 mm per year). During 2090 to 2099 under A1B, the central estimate of the rate of rise is 3.8 mm per year. For an average model, the scenario spread in sea-level rise is only 0.02 m by the middle of the century, and by the end of the century it is 0.15 m (Meehl *et al.*, 2007)". The projections of sea-level rise for the 12 SAP 2.1a scenarios by MAGICC are within the range reported by AR4 (Wigley *et al.*, 2007).

"Thermal expansion is the largest component, contributing 60 to 70 percent of the central estimate in these projections for all scenarios. Glaciers, ice caps and the Greenland ice sheet are also projected to contribute positively to sea level. GCMs indicate that the Antarctic ice sheet will receive increased snowfall without experiencing substantial surface melting, thus gaining mass and contributing negatively to sea level. Further accelerations in ice flow of the kind recently observed in some Greenland outlet glaciers and West Antarctic ice streams could substantially increase the contribution from the ice sheets. Current understanding of these effects is limited, so quantitative projections cannot be made with confidence (Meehl *et al.*, 2007)".

A.9 OCEAN CIRCULATION

a. There is no consistent change in the ENSO for those AOGCMs with a quasi-realistic base state.
b. Among those models with a realistic Atlantic Meridional Overturning Circulation (MOC), while it is very likely that the MOC will slow by 2100, there is little agreement among models for the magnitude of the slow-down. Models agree that the MOC will not shut down completely (Meehl *et al.*, 2007).

A.10 MONSOONS

Current AOGCMs predict that, in a warmer climate, there will be an increase in precipitation in both the Asian monsoon (along with an increase in interannual variability) and the southern part of the West African monsoon with some decrease in the Sahel in northern summer, as well as an increase in the Australian monsoon in southern summer. The monsoonal precipitation in Mexico and Central America is projected to decrease in association with increasing precipitation over the eastern equatorial Pacific. However, the uncertain role of particles complicates the projections of monsoon precipitation, particularly in the Asian monsoon (Meehl *et al.*, 2007).

A.11 TROPICAL CYCLONES (HURRICANES AND TYPHOONS)

The Summary for Policymakers finds it *likely* that intense hurricanes and typhoons will increase through the twenty-first century as it warms. Results from embedded high-resolution models and global models, ranging in grid spacing from 1 degree to 9 km, generally project increased peak wind intensities and notably, where analyzed, increased near-storm precipitation in future tropical cyclones (Meehl *et al.*, 2007). However, these questions of changes in frequency and intensity under global warming continue to be the subject of very active research (CCSP, 2008; Emanuel *et al.*, 2008; Knutson *et al.*, 2008).

A.12 MIDLATITUDE STORMS

Model projections show fewer midlatitude storms averaged over each hemisphere, associated with the poleward shift of the storm tracks that is particularly notable in the Southern Hemisphere, with lower central pressures for these poleward-shifted storms. The increased wind speeds result in more extreme wave heights in those regions (Meehl *et al.*, 2007).

A.13 RADIATIVE FORCING

"The radiative forcings by long-lived greenhouse gases computed with the radiative transfer codes in twenty of the AOGCMs used in the AR4 have been compared against results from benchmark line-by-line (LBL) models. The mean AOGCM forcing over the period 1860 to 2000 agrees with the mean LBL value to within 0.1 W per m^2 at the tropopause. However, there is a range of 25 percent in longwave forcing due to doubling CO_2 from its concentration in 1860 across the ensemble of AOGCM codes. There is a 47 percent relative range in longwave forcing at 2100 contributed by all greenhouse gases in the A1B scenario across the ensemble of AOGCM simulations. These results imply that the ranges in climate sensitivity and climate response from models discussed in this chapter may be due in part to differences in the formulation and treatment of radiative processes among the AOGCMs (Meehl *et al.*, 2007)".

A.14 CLIMATE CHANGE COMMITMENT (TEMPERATURE AND SEA LEVEL)

"Results from the AOGCM multi-model climate change commitment experiments (concentrations stabilized for 100 years at year 2000 for twentieth century commitment, and at 2100 values for B1 and A1B commitment) indicate that if greenhouse gases were stabilized, then a further warming of 0.5°C would occur (Meehl *et al.* 2007)".

"If concentrations were stabilized at A1B levels in 2100, sea-level rise due to thermal expansion in the twenty-second century would be similar to the twenty-first, and would amount to 0.3 to 0.8 m above present by 2300. The ranges of thermal expansion overlap substantially for stabilization at different levels, since model uncertainty is dominant; A1B is given here because most model results are available for that scenario. Thermal expansion would continue over many centuries at a gradually decreasing rate, reaching an eventual level of 0.2 to 0.6 m per degree of global warming relative to present (Meehl *et al.*, 2007)".

APPENDIX B

MAGICC Model Description
(Supplemental to Chapter 2)

Lead Author: Hiram Levy II, NOAA/GFDL

Contributing Author: Tom Wigley, NCAR

MAGICC (Model for the Assessment of Greenhouse-gas Induced Climate Change) is a coupled gas-cycle/climate model. Various versions of MAGICC have been used in all IPCC assessments. The version used here is the one that was used in the IPCC Third Assessment Report (TAR; IPCC, 2001; Cubasch and Meehl, 2001; Wigley and Raper, 2001). A critical assessment focused on its skill in predicting global average sea-level rise is found in Chapter 10 Appendix 1 of the Working Group I contribution to the Fourth Assessment Report (AR4) of the IPCC (Meehl *et al.,* 2007).

The climate component is an energy-balance model with a one-dimensional, upwelling-diffusion ocean (UDEBM). For further details of models of this type, see Hoffert *et al.* (1980) and Harvey *et al.* (1997). In MAGICC, the globe is divided into land and ocean "boxes" in both hemispheres in order to account for different thermal inertias and climate sensitivities over land and ocean, and hemispheric and land/ocean differences in forcing for short-lived gases and particles.

In order to allow inputs as emissions, the climate model is coupled interactively to a series of gas-cycle models for CO_2, CH_4, N_2O, a suite of halocarbons and SF_6. Details of the carbon cycle model are given in Wigley (1991a, 1993, 2000). The carbon cycle model includes both CO_2 fertilization and temperature feedbacks, with model parameters tuned to give results consistent with the other two carbon cycle models used in the TAR; *viz.* ISAM (Kheshgi and Jain, 2003) and the Bern model (Joos *et al.,* 2001) over a wide range of emissions scenarios. Details are given in Wigley

et al. (2007). The other gas cycle models are those used in the TAR (Prather and Ehhalt, 2001; Wigley *et al.,* 2002). Radiative forcings for the various gases are as used in the TAR. For sulfate particles, both direct and indirect forcings are included using forcing/emissions relationships developed in Wigley (1989, 1991b), with central estimates for 1990 forcing values. Sea-level rise estimates use thermal expansion values calculated directly from the climate model. Ice melt and other contributions are derived using formulae given in the TAR (Church and Gregory, 2001), except for the glacier and small ice cap contribution which employs an improved formulation that can be applied beyond 2100 (Wigley and Raper, 2005).

The standard inputs to MAGICC are emissions of the various radiatively important gases and various climate model parameters. For the TAR, these parameters were tuned so that MAGICC was able to emulate results from a range of Atmosphere-Ocean General Circulation Model (AOGCMs) (Cubasch and Meehl, 2001; Raper *et al.,* 2001). For the present calculations, a central set of parameters has been used. The most important of these is the climate sensitivity, where we have used a value of 2.6°C equilibrium global-mean warming for a CO_2 doubling, the median of values for AOGCMs used in the TAR.

Composition Models
(Supplemental to Chapter 2)

Lead Authors: Drew T. Shindell, NASA/GISS; M. Daniel Schwarzkopf, NOAA/GFDL; Larry W. Horowitz, NOAA/GFDL

Contributing Authors: Jean-Francois Lamarque, NCAR; Tom Wigley, NCAR

C.I GEOPHYSICAL FLUID DYNAMICS LABORATORY

Composition changes for the short-lived gases and particles from 2000 to 2100 in the GFDL experiments were calculated using the global chemical transport model MOZART-2 (Model for Ozone And Related chemical Tracers, version 2.4), which has been described in detail previously (Horowitz *et al.*, 2003; Horowitz, 2006; and references therein). This model was used to generate distributions of ozone, sulfate, black and organic carbon, and dust for the emissions scenarios discussed in Section 3.2.1. Emissions and initial conditions for methane were scaled each decade to match the global average methane abundances specified in the A1B "marker" scenario. The model includes 63 gases, 11 particles and precursor gases to simulate sulfate, nitrate, ammonium, and black and organic carbon, and five size bins for mineral dust (diameter size bins of 0.2 to 2.0 μm, 2.0-3.6 μm, 3.6 to 6.0 μm, 6.0 to 12.0 μm, and 12.0 to 20.0 μm). Hydrophobic black and organic carbon are chemically transformed into hydrophilic forms with a lifetime of 1.63 days (Tie *et al.*, 2005). Different particle types are assumed to be externally mixed and do not interact with one another. Sulfur oxidation in the gas phase and within clouds is fully interactive with the gas-phase oxidant chemistry.

The transport in MOZART-2 is driven with meteorological inputs provided every three hours by the middle-atmosphere version of the NCAR Community Climate Model (Kiehl *et al.*, 1998). The meteorology was the same for each decade, thus excluding any feedback from climate change on natural emissions and rates of chemical reactions and removal. Thus, natural emissions, such as those of isoprene, dust, and NO_x from lightning, are held constant at present-day levels. Convective mass fluxes are rediagnosed from the large-scale meteorology using the Hack (1994) and Zhang and McFarlane (1995) schemes. Vertical diffusion within the boundary layer is diagnosed using the scheme of Holtslag and Boville (1993). Tracer advection is performed using a flux-form semi-Lagrangian scheme (Lin and Rood, 1996).

The horizontal resolution is 2.8° latitude x 2.8° longitude, with 34 hybrid sigma-pressure levels extending up to 4 hPa. Photolysis frequencies for clear-sky are interpolated from a precalculated lookup table, based on a standard radiative transfer calculation (TUV version 3.0; Madronich and Flocke, 1998). The values are modified to account for cloudiness (Brasseur *et al.*, 1998), but do not account for effects of the simulated particles. Heterogeneous hydrolysis of N_2O_5 and NO_3 on particle surfaces occurs at a rate based on the simulated sulfate surface area, with a reaction probability of 0.04 (Tie *et al.*, 2005). Stratospheric concentrations of ozone and several other long-lived gases are relaxed to present-day climatological values.

Dry deposition velocities for reactive gases are calculated offline using a resistance-in-series scheme (Wesely, 1989). Deposition velocities for particles are prescribed as by Tie *et al.* (2005). Wet removal of soluble gases and particles in and below clouds is included as a first-order loss process, based on the large-scale and convective precipitation rates, as described by Horowitz *et al.* (2003). In-cloud scavenging is based on the parameterization of Giorgi and Chameides (1985), while below-cloud washout of highly soluble gases and particles follows Brasseur *et al.* (1998). For gases, the removal rate depends strongly on the temperature-dependent effective Henry's law constant. Wet deposition of soluble particles (sulfate, hydrophilic BC, hydrophilic OC, ammonium, and nitrate) is calculated by scaling the removal rate to that of highly-soluble HNO_3, assuming the particles have a first-order loss rate constant equal to 20 percent of that of HNO_3 (Tie *et al.*, 2005). This scaling introduces a large uncertainty into the calculation of particle burdens. Wet removal of dust is calculated using the formulation of Zender *et al.* (2003), with below-cloud scavenging efficiencies of 0.02 m^2 per kg for convective and 0.04 m^2 per kg for stratiform precipitation.

The ozone and particle distributions from these simulations have been evaluated by Horowitz (2006) and Ginoux *et al.* (2006), respectively. Simulated ozone concentrations agree well with present-day observations and recent trends (Horowitz, 2006). Overall, the predicted concentrations of particle are within a factor of two of the observed values and have a tendency to be overestimated (Ginoux *et al.*, 2006). The annual mean surface sulfate concentrations match observed values within a factor of two with values ranging from 0.05 μg per m^3 in the remote marine atmosphere to 13 μg per m^3 in polluted regions. In general, the simulated concentrations are over-predicted in summer and under-predicted in winter. Sulfate mass column and zonal mean profiles are comparable to previous studies, although the global mean burden is about 15 percent higher. The annual mean concentration of carbonaceous particles is generally overestimated in polluted regions by up to a factor of two. An exception is West Africa, where other models show significant loadings of carbonaceous particles associated with biomass burning activities during the dry season, while our results do not show any perturbation arising from such activities. The source of this discrepancy seems to be caused in part by the emission inventory in West Africa. The annual mean dust concentration at the surface agrees with the observations to within a factor of two, except over Antarctica where it is underestimated by a factor of five.

The three-dimensional monthly mean distributions of ozone, black and organic carbon particles, and sulfate particles from MOZART-2 were archived from simulations for each decade from 2000 to 2100. The results from these simulations were then interpolated to intermediate years and used in the transient climate simulations. The distribution of dust from a present-day simulation in MOZART-2 was used in all years of the climate simulations.

C.2 GODDARD INSTITUTE FOR SPACE STUDIES

The configuration of the GISS composition model used here has been described in detail in (Shindell *et al.*, 2007). The composition model PUCCINI (Physical Understanding of Composition-Climate INteractions and Impacts) includes ozone and oxidant photochemistry in both the troposphere and stratosphere (Shindell *et al.*, 2006). Photochemistry includes 155 reactions. The model calculates the abundances of 51 chemicals, 26 of which are transported by the model's advection scheme. It uses "lumped families" for hydrocarbons and PANs. Chemical reactions involving these surrogates are based on the similarity between the molecular bond structures within each family using the reduced chemical mechanism of (Houweling *et al.*, 1998). This mechanism is based on the Carbon Bond Mechanism-4 (CBM-4) (Gery *et al.*, 1989), modified to better represent the globally important range of conditions. The CBM-4 scheme has been validated extensively against smog chamber experiments and more detailed chemical schemes. This scheme was modified for use in global models by removing aromatic compounds and adding in reactions important in background conditions, including organic nitrate and organic peroxide reactions, and extending the methane oxidation chemistry. The revised scheme was then readjusted based on the more extensive Regional Atmospheric Chemistry Model (RACM) (Stockwell *et al.*, 1997), and the modified scheme includes several surrogates designed to compensate for biases relative to the RACM mechanism. The modified scheme was shown to agree well with the detailed RACM reference mechanism over a wide range of chemical conditions including relatively pristine environments (Houweling *et al.*, 1998).

Rate coefficients are taken from the NASA JPL 2000 handbook (Sander *et al.*, 2000). Photolysis rates are calculated using the Fast-J2 scheme (Bian and Prather, 2002), except for the photolysis of water and nitric oxide (NO) in the Schumann-Runge bands, which are parameterized according to Nicolet (1984; Nicolet and Cieslik, 1980). The particle component simulates sulfate, carbonaceous and sea salt particles (Koch *et al.*, 2007; Koch *et al.*, 2006) and nitrate particles (Bauer *et al.*, 2007). It includes prognostic simulations of DMS, MSA, SO_2 and sulfate mass distributions. The mineral dust particle model transports four different sizes classes of dust particles with radii between 0.1 to 1, 1 to 2, 2 to 4, and 4 to 8 microns (Miller *et al.*, 2006). Most importantly, these components interact with one another, with linkages including oxidants affecting sulfate, gases affecting nitrate, sulfate affecting nitrogen heterogeneous chemistry via reaction of N_2O_5 to HNO_3, and sulfate and nitrate being absorbed onto mineral dust surfaces (*i.e.*, the particles are internally mixed as coatings form on dust surfaces (Bauer *et al.*, 2007). The latter is described by a pseudo first-order rate coefficient which gives the net irreversible removal rate of gases by an particle surface. We use the uptake coefficient of 0.1 recommended from laboratory measurements (Hanisch and Crowley, 2001), though this value is fairly uncertain.

Phase transformation and removal is calculated using a wet deposition scheme in which soluble gases can be removed into either moist convective plumes or large-scale clouds as derived from the GCM's internal cloud scheme (Del Genio and Yao, 1993). During convection, all gases and particles are transported along with the convective plumes, with scavenging of soluble gases and particles within and below cloud updrafts. In large-scale stratiform clouds, soluble gases are removed based on the fraction of the grid box over which precipitation is taking place. Washout of soluble gases and particles is calculated below precipitating clouds. In the case of either evaporation of precipitation before reaching the ground, or detrainment or evaporation from a convective updraft, the dissolved gases and particles are returned to the air. Wet chemistry calculations take place in each grid box at each time step, including the coupling with the convection scheme's entraining and nonentraining plumes (which are based on the convective instability in the particular grid box at that time), so are entirely consistent with the contemporaneous model physics. The solubility of each gas is determined by an effective Henry's Law coefficient, assuming a pH of 4.5. Surface dry deposition is calculated using a resistance-in-series model (Wesely and Hicks, 1977) coupled with a global, seasonally varying vegetation data set as given by Chin *et al.* (1996).

The 2000 simulation uses the 2000 emission inventory of the International Institute for Applied Systems Analysis (IIASA), except for biomass burning which is taken from the Global Fire Emission Database (GFED) averaged over 1997 to 2002 (Van der Werf *et al.*, 2003) with emission factors from (Andrae and Merlet, 2001) for particles. The IIASA inventory is based on the 1995 EDGAR3.2 inventory (Olivier and Berdowski, 2001), extrapolated to 2000 using national and sector economic development data (Dentener *et al.*, 2005). Lightning NO_x emissions are calculated internally in the GCM (5.6 Tg per year for present-day), and other natural sources are prescribed according to conventional estimates. Dust emissions are constant at 1580 Tg per year, while isoprene emissions are 356 Tg per year. Emissions of DMS are 41 Tg per year.

The simulations described here were run with this composition model included within a 23-layer (up to 0.01 hPa), 4° x 5° horizontal resolution version of the ModelE climate model (Schmidt *et al.*, 2006). This composition model was used for both the transient climate and regional/sector emissions perturbation simulations.

Present-day composition results in the model are generally similar to those in the underlying chemistry and particle models documented previously. The model used here does not include the enhanced convective scavenging of insoluble gases and particles prescribed in Koch *et al.* (2007). Therefore our carbonaceous particle burden, especially in the free troposphere, is nearly double that of Koch *et al.* (2007). Comparison with the limited available observations is comparable between the two simulations (a positive bias replaces a negative bias).

C.3 NATIONAL CENTER FOR ATMOSPHERIC RESEARCH

Various methods were used at NCAR to estimate future composition. Present-day tropospheric ozone was taken from calculations performed by Lamarque *et al.* (2005) using the MOZART-2 model; beyond 2000, tropospheric ozone was calculated by T. Wigley using the MAGICC model <http://www.cru.uea.ac.uk/~mikeh/software/magicc.htm> forced by the time-varying emissions of NO_x, methane and VOCs and these average global values were used to scale the present-day distribution. Future carbonaceous particles were scaled from their present-day distribution (Collins *et al.*, 2001) by a globally uniform factor whose time evolution follows the global evolution of SO_2 emissions. Future levels of sulfate particles were calculated using the MOZART model. Stratospheric ozone changes are prescribed following the study by (Kiehl *et al.*, 1999).

The Model for Ozone and Related chemical Tracers version 2 (MOZART-2) is described by Horowitz *et al.* (2003) and references therein. The model provides the distribution of 80 chemical constituents (including nonmethane hydrocarbons) between the surface and the stratosphere. The model was run at a uniform horizontal resolution of ~2.8° in both latitude and longitude. The vertical discretization of the meteorological data (described below) and hence of the model consists of 18 hybrid levels from the ground to ~4 hPa. The evolution of gases and particles is calculated with a time step of 20 minutes.

The tropospheric photolysis rates use a vertical distribution of ozone based on the simulated ozone in the troposphere and on the climatology from Kiehl *et al.* (1999) above. For each simulation, this latter distribution is updated to reflect the changes in the lower stratosphere during the twentieth century, affecting only the photolysis rates and not the amount of ozone transported from the stratosphere.

The NCAR regional/sector perturbation simulations (Section 3.4) used a version of MOZART chemical transport model (Horowitz *et al.*, 2003) embedded within the Community Atmosphere Model (CAM3, Collins *et al.*, 2006). This model, known as CAM-chem, includes an extension of the chemical mechanism presented by Horowitz *et al.* (2003) to include an updated terpene oxidation scheme and a better treatment of anthropogenic non-methane hydrocarbons (NMHCs). The MOZART particles have been extended by Tie *et al.* (2001, 2005) to include a representation of ammonium nitrate that is dependent on the amount of sulfate and ammonia present in the air mass following the parameterization of gas/particle partitioning by Metzger *et al.* (2002). In brief, CAM-chem simulates the evolution of the bulk particle mass of black carbon (BC, hydrophobic and hydrophilic), primary organic (POA, hydrophobic and hydrophilic), second organic (SOA, linked to the gas-phase chemistry through the oxidation of atmospheric NMHCs as in (Chung and Seinfeld, 2002), ammonium and ammonium nitrate (from NH_3 emissions), and sulfate particles (from SO_2 and DMS emissions). It also considers the uptake of N_2O_5, HO_2, NO_2, and NO_3 on particles. Results from the CAM-chem model are discussed by Lamarque *et al.* (2005). A description of sea salt, updated from Tie *et al.* (2005), is also included. Finally, a monthly-varying climatology of dust is used only for radiative calculations. The CAM-chem model considers only the direct effect of particles and the atmospheric model is coupled with the chemistry solely through the radiative fluxes, taking into account all radiatively active gases and particles. The horizontal resolution is 2° latitude x 2.5° longitude, with 26 levels ranging from the surface to ~4 hPa.

APPENDIX D

Climate Models
(Supplemental to Chapter 3)

Lead Authors: Drew T. Shindell, NASA/GISS; M. Daniel Schwarzkopf, NOAA/GFDL; Hiram Levy II, NOAA/GFDL; Larry W. Horowitz, NOAA/GFDL

Contributing Author: Jean-Francois Lamarque, NCAR

D.1 GEOPHYSICAL FLUID DYNAMICS LABORATORY

Climate simulations at GFDL used the coupled climate model recently developed at NOAA's Geophysical Fluid Dynamics Laboratory, which has been previously described in detail (Delworth *et al.*, 2006). A brief summary is provided here. The model simulates atmospheric and oceanic climate and variability from the diurnal time-scale through multi-century climate change without employing flux adjustment. The control simulation has a stable, realistic climate when integrated over multiple centuries and a realistic ENSO (Wittenberg *et al.*, 2006). Its equilibrium climate response to a doubling of CO_2 is 3.4°C (Stouffer *et al.*, 2006). There are no indirect particle effects included in any of the simulations. The resolution of the land and atmospheric components is 2.5° longitude x 2° latitude and the atmospheric model has 24 vertical levels. The ocean resolution is 1° latitude x 1° longitude, with meridional resolution equatorward of 30° becoming progressively finer, such that the meridional resolution is 1/3° at the Equator. There are 50 vertical levels in the ocean, with 22 evenly-spaced levels within the top 220 m. The ocean component has poles over North America and Eurasia to avoid polar filtering.

Using a five-member ensemble simulation of the historical climate (1861 to 2003), including the evolution of natural and anthropogenic forcing agents, the GFDL climate model is able to capture the global historical trend in observed surface temperature for the twentieth century as well as many continental-scale features (Knutson *et al.*, 2006). However, the model shows some tendency for too much twentieth century warming in lower latitudes and too little warming in higher latitudes. Differences in Arctic Oscillation behavior between models and observations contribute substantially to an underprediction of the observed warming over northern Asia. El Niño interactions complicate comparisons of observed and simulated temperature records for the El Chichón and Mt. Pinatubo eruptions during the early 1980s and early 1990s (Knutson *et al.*, 2006). In Figure 7d of Knutson *et al.* (2006), where the model ensemble and observations are compared grid box by grid box, ~60 percent of those grid boxes with sufficient observational data have twentieth century surface temperature trends that agree quantitatively with the model ensemble. In general, many observed continental-scale features, including a twentieth century cooling over the North Atlantic, are captured by the model ensemble, as Figures 7a and 7c in Knutson *et al.* (2006) show. However, the model ensemble does not capture the observed cooling over the southeastern United States, and it produces a twentieth century cooling over the North Pacific that is not observed.

D.2 GODDARD INSTITUTE FOR SPACE STUDIES

The GISS climate simulations were performed using GISS ModelE (Schmidt *et al.*, 2006). We use a 20-layer version of the atmospheric model (up to 0.1 hPa) coupled with a dynamic ocean without flux adjustment, both run at four by five degree horizontal resolution, as in the GISS-ER IPCC AR4 simulations (Hansen *et al.*, 2007). This model has been extensively evaluated against observations (Schmidt *et al.*, 2006), and has a climate sensitiv-

ity in accord with values inferred from paleoclimate data and similar to that of mainstream GCMs: an equilibrium climate sensitivity of 2.6°C for doubled CO_2.

The modeled radiatively active gases and particles influence the climate in the GCM. Ozone and particles can affect both the short and long wavelength radiation flux. Water uptake on particle surfaces influences the particle effective radius, refractive index and extinction efficiency as a function of wavelength and the local relative humidity (Koch *et al.*, 2007), which in turn affects the GCM's radiation field.

The GISS model also includes a simple parameterization for the particle indirect effect (Menon *et al.*, 2002) (Box 3.1). For the present simulations, we use only cloud cover changes (the second indirect effect), with empirical coefficients selected to give roughly -1 W per m^2 forcing from the preindustrial era to the present, a value chosen to match diurnal temperature and satellite polarization measurements, as described in Hansen *et al.* (2005). We note, however, that this forcing is roughly twice the value of many other model studies (Penner *et al.*, 2006). The particle indirect effect in the model takes place only from the surface through ~570 hPa, as we only let particles affect liquid-phase stratus clouds.

D.3 NATIONAL CENTER FOR ATMOSPHERIC RESEARCH

The transient climate simulations use the NCAR Community Climate System Model CCSM3 (Collins *et al.*, 2006). This model had been run previously with evolution of short-lived gases and particles in the future for the IPCC AR4. The model was run at T85 (~1.4° x 1.4° resolution). For this study, a new simulation was performed for 2000 to 2050 in which ozone and particles were kept at their 2000 levels. The equilibrium climate sensitivity of this model to doubled CO_2 is 2.7°C.

APPENDIX E

Scenarios
(Supplemental to Chapter 4)

Lead Authors: Drew T. Shindell, NASA/GISS; Hiram Levy II, NOAA/GFDL

Contributing Author: Anne Waple, STG Inc.

E.1 THE EMISSIONS SCENARIOS OF THE IPCC SPECIAL REPORT ON EMISSIONS SCENARIOS (SRES) (NAKIĆENOVIĆ AND SWART, 2000)

A1. The A1 storyline and scenario family describes a future world of very rapid economic growth, global population that peaks in mid-century and declines thereafter, and the rapid introduction of new and more efficient technologies.

Major underlying themes are convergent among regions, capacity building, and increased cultural and social interactions, with a substantial reduction in regional differences in per capita income. The A1 scenario family develops into three groups that describe alternative directions of technological change in the energy system. The three A1 groups are distinguished by their technological emphasis: fossil intensive (A1FI), non-fossil energy sources (A1T), or a balance across all sources (A1B) (where balanced is defined as not relying too heavily on one particular energy source, on the assumption that similar improvement rates apply to all energy supply and end-use technologies).

A2. The A2 storyline and scenario family describes a very heterogeneous world. The underlying theme is self reliance and preservation of local identities. Fertility patterns across regions converge very slowly, which results in continuously increasing population. Economic development is primarily region-

ally oriented, and per capita economic growth and technological change more fragmented and slower than other storylines.

B1. The B1 storyline and scenario family describes a convergent world with the same global population as in the A1 storyline (one that peaks in mid-century and declines thereafter), but with rapid change in economic structures toward a service and information economy, with reductions in material intensity and the introduction of clean and resource-efficient technologies. The emphasis is on global solutions to economic, social, and environmental sustainability, including improved equity, but without additional climate initiatives.

B2. The B2 storyline and scenario family describes a world in which the emphasis is on local solutions to economic, social, and environmental sustainability. It is a world with continuously increasing global population, at a rate lower than for the A2 storyline, intermediate levels of economic development, and less rapid and more diverse technological change than in the B1 and A1 storylines. While the scenario is also oriented toward environmental protection and social equity, it focuses on local and regional levels.

An illustrative scenario was chosen for each of the six scenario groups A1B, A1FI, A1T, A2, B1 and B2. All should be considered equally sound.

The SRES scenarios do not include additional climate initiatives, which means that no scenarios are included that explicitly assume implementation of the United Nations Framework Convention on Climate Change or the emissions targets of the Kyoto Protocol.

E.2 RADIATIVE FORCING STABILIZATION LEVELS AND APPROXIMATE CO$_2$ CONCENTRATIONS FROM THE CCSP SAP 2.1A SCENARIOS (TABLE 1.2; CLARKE ET AL., 2007)

The stabilization levels were constructed so that the CO$_2$ concentrations resulting from stabilization of total radiative forcing in Watts per square meter (W per m^2), after accounting for radiative forcing from the non-CO$_2$ greenhouse gases (GHGs) included in this research, would be roughly 450 parts per million by volume (ppmv), 550 ppmv, 650 ppmv, and 750 ppmv.

	Total Radiative Forcing from GHGs (W per m^2)	Approximate Contribution to Radiative Forcing from Non-CO$_2$ GHGs (W per m^2)	Approximate Contribution to Radiative Forcing from CO$_2$ (W per m^2)	Corresponding CO$_2$ Concentration (ppmv)
Level 1	3.4	0.8	2.6	450
Level 2	4.7	1.0	3.7	550
Level 3	5.8	1.3	4.5	650
Level 4	6.7	1.4	5.3	750
Year 1998	2.11	0.65	1.46	365
Preindustrial	0	0	0	275

GLOSSARY AND ACRONYMS

GLOSSARY

aerosols
very small airborne solid or liquid particles that reside
in the atmosphere for at least several hours, with the
smallest remaining airborne for days

anthropogenic
human-induced

attribution
attribution of causes of climate change is the process
of establishing the most likely causes for a detected
change with some defined level of confidence

black carbon
soot particles primarily from fossil fuel burning

climate sensitivity
the equilibrium change in global-average surface air
temperature following a change in radiative forcing;
in current usage, this term generally refers to the
warming that would result if atmospheric carbon
dioxide concentrations were to double from their
pre-industrial levels

cyclone
a storm system that rotates around a center of low
atmospheric pressure

tropical cyclone
a cyclone usually originating in the tropics, with a
warm central core

extratropical cyclone
a cyclone originating in the mid or high latitudes,
with a cold central core. Larger in scale than a tropical
cyclone and with less central intensity

diurnal temperature range
the difference between maximum and minimum
temperature over a period of 24 hours

El Niño-Southern Oscillation (ENSO)
the waxing and waning every two to seven years of
El Niño and La Niña ocean temperature cycles along
with the related atmospheric pressure component of
the Southern Oscillation; the primary centers of ENSO
variability are in the tropical Pacific, but ENSO effects
can be felt across much of the globe

forcing
a nature or human-induced factor that influences climate

greenhouse gases
gases including water vapor, carbon dioxide, methane,
nitrous oxide, and halocarbons that trap infrared heat,
warming the air near the surface and in the lower levels
of the atmosphere

halocarbons
chemical compounds that contain at least one carbon atom
and at least one halogen atom, such as chlorine, fluorine,
bromine, or iodine

inhomogeneity
a break or interruption in an otherwise homogeneous
record; for example, moving a weather station from the
center of a city to the suburbs will create an inhomogeneity in the climate record

monsoon
a seasonal change in wind direction (driven by changes in
temperature), often accompanied by a seasonal precipitation maximum

organic carbon
particles consisting predominantly of organic compounds,
mainly carbon, hydrogen, oxygen and lesser amounts of
other elements

parameterization
a mathematical representation of a process that cannot be
explicitly resolved in a climate model

stratosphere
the highly stratified region of the atmosphere above the
troposphere extending from about 10 kilometers (km)
(ranging from 9 km in high latitudes to 16 km in the
tropics on average) to about 50 km

troposphere
the lowest part of the atmosphere from the surface to
about 10 km in altitude in mid-latitudes (ranging from 9
km in high latitudes to 16 km in the tropics on average)
where clouds and "weather" phenomena occur, in the troposphere, temperatures generally decrease with height

ACRONYMS

AERONET	Aerosol Robotic Network
AIM	Asian-Pacific Integrated Model
AOD	Aerosol Optical Depth
AOGCM	Atmosphere-Ocean (-Sea Ice) General Circulation Model
AR4	IPCC Fourth Assessment Report
ARL	Air Resources Laboratory
AVHRR	Advanced Very High Resolution Radiometer
BC	black carbon
CaCO$_3$	calcium carbonate
CAM3	Community Atmosphere Model
CCSM	Community Climate System Model
CCSP	Climate Change Science Program
CH$_4$	methane
CO$_2$	carbon dioxide
DMS	dimethylsulfide
ENSO	El Niño-Southern Oscillation
EPA	Environmental Protection Agency
ESM	Earth System Model
FAR	First IPCC Assessment Report
GCM	General Circulation Model/Global Climate Model
GFDL	Geophysical Fluid Dynamics Laboratory
GHG	greenhouse gas
GISS	Goddard Institute for Space Studies
hPa	hectopascal
HNO$_3$	nitric acid
HO$_2$	hydroperoxyl radical
H$_2$O$_2$	hydrogen Peroxide
IAM	Integrated Assessment Model
IGSM	MIT Integrated Global System Model
IMAGE	Integrated Model to Assess the Greenhouse Effect
IPCC	Intergovernmental Panel on Climate Change
LL	long-lived
mW per m^2	milliwatts per square meter (after MAGICC)
MACCM3	NCAR Middle Atmosphere Community Climate Model, version 3
MAGICC	Model for the Assessment of Greenhouse-gas Induced Climate Change
MODIS	Moderate Resolution Imaging Spectroradiometer
MOZART	Model for Ozone and Related Chemical Tracers
NASA	National Aeronautics and Space Administration
NCAR	National Center for Atmospheric Research
NCDC	National Climatic Data Center
NH	Northern Hemisphere
NH$_3$	ammonia
NMHC	non-methane hydrocarbons
N$_2$O	nitrous oxide
N$_2$O$_5$	nitric pentoxide
NOAA	National Oceanic and Atmospheric Administration
NO$_2$	nitrogen dioxide
NO$_3$	nitrate radical
NO$_x$	reactive nitrogen oxides
NRC	National Research Council
NSF	National Science Foundation
O$_3$	ozone
OC	organic carbon
ppbv	parts per billion by volume
ppm	parts per million
ppmv	parts per million by volume
pptv	parts per trillion by volume
RF	radiative forcing
SAP	Synthesis and Assessment Product
SAR	IPCC Second Assessment Report
SH	Southern Hemisphere
SL	short-lived
SO$_2$	sulfur dioxide
SO$_4$	sulfate
SOA	secondary organic aerosol
SRES	Special Report on Emissions Scenarios
SST	sea surface temperature
TAR	Third IPCC Assessment Report
Tg	teragrams, $1 Tg = 1 \times 10^{12}$ grams
TUV	Tropospheric Ultraviolet and Visible Radiation Model
VOCs	Volatile Organic Compounds
W per m^2	watts per square meter

REFERENCES

PREFACE REFERENCES

Clarke, L., J. Edmonds, H. Jacoby, H. Pitcher, J. Reilly, and R. Richels, 2007: *Scenarios of Greenhouse Gas Emissions and Atmospheric Concentrations.* Subreport 2.1A of Synthesis and Assessment Product 2.1 by the U.S. Climate Change Science Program and the Subcommittee on Global Change Research. Department of Energy, Office of Biological & Environmental Research, Washington, DC, 154 pp.

IPCC (Intergovernmental Panel on Climate Change) 2007: *Climate Change 2007: The Physical Science Basis.* Contribution of Working Group I to the Fourth Assessment Report (AR4) of the Intergovernmental Panel on Climate Change [Solomon, S., D. Qin, M. Manning, Z. Chen, M. Marquis, K.B. Averyt, M.Tignor, and H.L. Miller (eds.)]. Cambridge University Press, Cambridge, UK, and New York, 996 pp.

CHAPTER 1 REFERENCES

Brasseur, G.P. and E. Roeckner, 2005: Impact of improved air quality on the future evolution of climate. *Geophysical Research Letters,* **32(23)**, L23704, doi:10.1029/2005GL023902.

Clarke, L., J. Edmonds, H. Jacoby, H. Pitcher, J. Reilly, and R. Richels, 2007: *Scenarios of Greenhouse Gas Emissions and Atmospheric Concentrations.* Subreport 2.1A of Synthesis and Assessment Product 2.1 by the U.S. Climate Change Science Program and the Subcommittee on Global Change Research. Department of Energy, Office of Biological & Environmental Research, Washington, DC, 154 pp.

Delworth, T.L., V. Ramaswamy, and G.L. Stenchikov, 2005: The impact of aerosols on simulated ocean temperature and heat content in the 20th century. *Geophysical Research Letters,* **32(24)**, L24709, doi:10.1029/2005GL024457.

Hansen, J., A. Lacis, D. Rind, G. Russell, P. Stone, I. Fung, R. Ruedy, and J. Lerner, 1984: Climate sensitivity: analysis of feedback mechanisms. In: *Climate Processes and Climate Sensitivity,* [Hansen, J.E. and T. Takahashi (eds.)]. AGU Geophysical Monograph 29, Maurice Ewing Vol. 5. American Geophysical Union, Washington DC, pp. 130-163.

Hansen, J., Mki. Sato, R. Ruedy, A. Lacis, and V. Oinas, 2000: Global warming in the twenty-first century: an alternative scenario. *Proceedings of the National Academy of Sciences,* **97(18)**, 9875-9880.

IPCC (Intergovernmental Panel on Climate Change), 1990: *Climate Change: The IPCC Scientific Assessment.* [Houghton, J.T., G.J. Jenkins, and J.J. Ephraums (eds.)]. Cambridge University Press, Cambridge, UK, and New York, 365 pp.

IPCC (Intergovernmental Panel on Climate Change), 1992: *Climate Change 1992: The Supplementary Report to the IPCC Scientific Assessment.* [Houghton, J.T., B.A. Callander, and S.K. Varney (eds.)]. Cambridge University Press, Cambridge, UK, and New York, 200 pp.

IPCC (Intergovernmental Panel on Climate Change), 1996: *Climate Change 1995: The Science of Climate Change.* Contribution of Working Group I to the Second Assessment Report of the Intergovernmental Panel on Climate Change. [Houghton, J.T, L.G. Meiro Filho, B.A. Callander, N. Harris, A. Kattenberg, and K. Maskell (eds.)]. Cambridge University Press, Cambridge, UK, and New York, 572 pp.

IPCC (Intergovernmental Panel on Climate Change), 2001: *Climate Change 2001: The Scientific Basis.* Contribution of Working Group I to the Third Assessment Report of the Intergovernmental Panel on Climate Change [Houghton, J. T., Y. Ding, D.J. Griggs, M. Noguer, P.J. van der Linden, X. Dai, K. Maskell, and C.A. Johnson (eds.)]. Cambridge University Press, Cambridge, UK, and New York, 881 pp.

IPCC (Intergovernmental Panel on Climate Change) 2007: *Climate Change 2007: The Physical Science Basis.* Contribution of Working Group I to the Fourth Assessment Report (AR4) of the Intergovernmental Panel on Climate Change [Solomon, S., D. Qin, M. Manning, Z. Chen, M. Marquis, K.B. Averyt, M.Tignor, and H.L. Miller (eds.)]. Cambridge University Press, Cambridge, UK, and New York, 996 pp.

Manabe, S. and R.J. Stouffer, 1979: A CO_2-climate sensitivity study with a mathematical model of the global climate. *Nature,* **282(5738)**, 491-493.

Manabe, S. and R.T. Wetherald, 1967: Thermal equilibrium of the atmosphere with a given distribution of relative humidity. *Journal of the Atmospheric Sciences,* **24(3)**, 241-259.

Mitchell, J.F.B., T.C. Johns, J.M. Gregory, and S.F.B. Tett, 1995: Climate response to increasing levels of greenhouse gases and sulphate aerosols. *Nature,* **376(6540),** 501-504.

Nakićenović, N. and R. Swart (eds.), 2000: *Special Report on Emissions Scenarios*. A special report of Working Group III of the Intergovernmental Panel on Climate Change. Cambridge University Press, Cambridge, UK, and New York, 599 pp.

Stouffer, R.J., S. Manabe, and K. Bryan, 1989: Interhemispheric asymmetry in climate response to a gradual increase of atmospheric CO_2. *Nature*, **342(3250)**, 660-662.

Washington, W.M. and G.A Meehl, 1989: Climate sensitivity due to increased CO_2: experiments with a coupled atmosphere and ocean general circulation model. *Geophysical Research Letters*, **4(1)**, 1-38.

CHAPTER 2 REFERENCES

Christensen, J.H., B. Hewitson, A. Busuloc, A. Chen, X. Gao, I. Held, R. Jones, R.K. Kolli, W.-T. Kwon, R. Laprise, V. Magaña Rueda, L. Mearns, C.G. Menéndez, J. Räisänen, A. Rinke, A. Sarr, and P. Whetton, 2007: Regional climate projections. In: *Climate Change 2007: The Physical Basis*. Contribution of Working Group I to the Fourth Assessment Report (AR4) of the Intergovernmental Panel on Climate Change [Solomon, S., D. Qin, M. Manning, Z. Chen, M. Marquis, K.B. Averyt, M. Tignor, and H.L. Miller (eds.)]. Cambridge University Press, Cambridge, UK, and New York, pp. 847-940.

Clarke, L., J. Edmonds, H. Jacoby, H. Pitcher, J. Reilly, and R. Richels, 2007: *Scenarios of Greenhouse Gas Emissions and Atmospheric Concentrations*. Sub-report 2.1A of Synthesis and Assessment Product 2.1 by the U.S. Climate Change Science Program and the Subcommittee on Global Change Research. Department of Energy, Office of Biological & Environmental Research, Washington, DC, 154 pp.

Cubasch, U. and G. A. Meehl, 2001: Projections for future climate change. In: *Climate Change 2001: The Scientific Basis*. Contribution of Working Group I to the Third Assessment Report of the Intergovernmental Panel on Climate Change [Houghton, J.T., Y. Ding, D.J. Griggs, M. Noguer, P.J. van der Linden, X. Dai, K. Maskell, and C.A. Johnson (eds.)]. Cambridge University Press, Cambridge, UK and New York, pp. 525-582.

Harvey, L.D.D., J. Gregory, M. Hoffert, A. Jain, M. Lal, R. Leemans, S.B.C. Raper, T.M.L. Wigley, and J. de Wolde, 1997: *An introduction to simple climate models used in the IPCC Second Assessment Report*. IPCC Technical Paper 2 [Houghton, J.T., L.G. Meira Filho, D.J. Griggs, and M. Noguer (eds.)]. Intergovernmental Panel on Climate Change, Geneva, Switzerland, 50 pp.

Hoffert, M.L., A.J. Callegari, and C.-T. Hsieh, 1980: The role of deep sea heat storage in the secular response to climate forcing. *Journal of Geophysical Research*, **86**, 6667- 6679.

IPCC (Intergovernmental Panel on Climate Change), 2001: *Climate Change 2001: Synthesis Report*. A Contribution of Working Groups I, II, and III to the Third Assessment Report of the Intergovernmental Panel on Climate Change [Watson, R.T. and the Core Writing Team (eds.)]. Cambridge University Press, Cambridge, UK, and New York, 398 pp.

IPCC (Intergovernmental Panel on Climate Change), 2007a: *Climate Change 2007: The Physical Science Basis*. Contribution of Working Group I to the Fourth Assessment Report (AR4) of the Intergovernmental Panel on Climate Change [Solomon, S., D. Qin, M. Manning, Z. Chen, M. Marquis, K.B.Averyt, M. Tignor and H.L. Miller (eds.)]. Cambridge University Press, Cambridge, UK and New York, 996 pp.

IPCC (Intergovernmental Panel on Climate Change), 2007b: Summary for policymakers. In: *Climate Change 2007: The Physical Science Basis*. Contribution of Working Group I to the Fourth Assessment Report (AR4) of the Intergovernmental Panel on Climate Change [Solomon, S., D. Qin, M. Manning, Z. Chen, M. Marquis, K.B. Averyt, M.Tignor, and H.L. Miller (eds.)]. Cambridge University Press, Cambridge, UK, and New York, pp. 1-18.

Joos, F., I.C. Prentice, S. Sitch, R. Meyer, G. Hooss, G.-K. Plattner, S. Gerber, and K. Hasselmann, 2001: Global warming feedbacks on terrestrial carbon uptake under the Intergovernmental Panel on Climate Change (IPCC) emissions scenarios. *Global Biogeochemical Cycles*, **15(4)**, 891-908, doi:10.1029/2000GB001375.

Kheshgi, H.S. and A.K. Jain, 2003: Projecting future climate change: implications of carbon cycle model intercomparisons. *Global Biogeochemical Cycles*, **17**, 1047, doi:10.1029/2001GB001842.

Kim, S.H., J.A. Edmonds, J. Lurz, S.J. Smith, and M.A. Wise, 2006: The ObjECTS framework for integrated assessment: Hybrid modeling of transportation. *The Energy Journal*, **27**(Special Issue No. 2), 51-80.

Manne, A.S. and R.G. Richels, 2001: An alternative approach to establishing trade-offs among greenhouse gases. *Nature*, **410(6829)**, 675-677.

Meehl, G.A., T.F. Stocker, W.D. Collins, P. Friedlingstein, A.T. Gaye, J.M. Gregory, A. Kitoh, R. Knutti, J.M. Murphy, A. Noda, S.C.B. Raper, I.G. Watterson, A.J. Weaver, and Z.-C. Zhao, 2007: Global climate projections. In: *Climate Change 2007: The Physical Basis*. Contribution of Working Group I to the Fourth Assessment Report (AR4) of the Intergovernmental Panel on Climate Change [Solomon, S., D. Qin, M. Manning, Z. Chen, M. Marquis, K.B. Averyt, M. Tignor, and H.L. Miller (eds.)]. Cambridge University Press, Cambridge, UK, and New York, pp. 747-845.

Nakićenović, N. and R. Swart (eds.), 2000: *Special Report on Emissions Scenarios.* A special report of Working Group III of the Intergovernmental Panel on Climate Change. Cambridge University Press, Cambridge, UK, and New York, 599 pp.

Paltsev, S., J.M. Reilly, H.D. Jacoby, R.S. Eckaus, J. McFarland, M.C. Sarofim, M. Asadoorian, and M. Babiker, 2005: *The MIT Emissions Prediction and Policy Analysis (EPPA) Model: Version 4.* Report no. 125. MIT Joint Program on the Science and Policy of Global Change, Cambridge, MA, 72 pp.

Reilly, J.M., R. Prinn, J. Harnisch, J. Fitzmaurice, H. Jacoby, D. Kicklighter, J. Melillo, P. Stone, I. Sokolov, and C. Wang, 1999: Multi-gas assessment of the Kyoto Protocol. *Nature,* **401(6753)**, 549-555.

Richels, R., A.S. Manne, and T.M.L. Wigley, 2007: Moving beyond concentrations: the challenge of limiting temperature change. In: *Human Induced Climate Change: An Interdisciplinary Assessment* [Schlesinger, M., F.C. de la Chesnaye, H. Kheshgi, C.D. Kolstad, J. Reilly, J.B. Smith, and T. Wilson (eds.)]. Cambridge University Press, Cambridge, UK, pp. 387-402.

Sarofim, M.C., C.E. Forest, D.M. Reiner, and J.M. Reilly, 2005: Stabilization and global climate policy. *Global and Planetary Change,* **47(2-4)**, 266-272.

Wigley, T.M.L., 1989: Possible climatic change due to SO_2-derived cloud condensation nuclei. *Nature,* **339(6223)**, 365-367.

Wigley, T.M.L., 1991: Could reducing fossil-fuel emissions cause global warming? *Nature,* **349(6309)**, 503-506.

Wigley, T.M.L. and S.C.B. Raper, 2001: Interpretation of high projections for global-mean warming. *Science,* **293(5529)**, 451-454.

Wigley, T.M.L., R. Richels, and J.A. Edmonds, 1996: Economic and environmental choices in the stabilization of atmospheric CO_2 concentrations. *Nature,* **379(6562)**, 240-243.

Wigley, T.M.L., L.E. Clarke, J.A. Edmonds, H.D. Jacoby, S. Paltsev, H. Pitcher, J. Reilly, R. Richels, M.C. Sarofim, and S.J. Smith, 2008: Uncertainties in climate stabilization. *Climatic Change,* in press.

CHAPTER 3 REFERENCES

Albrecht, B.A., 1989: Aerosols, cloud microphysics, and fractional cloudiness. *Science,* **245(4923)**, 1227-1230.

Andreae, M.O. and P. Merlet, 2001: Emission of trace gases and aerosols from biomass burning. *Global Biogeochemical Cycles,* **15(4)**, 955-966.

Bauer, S.E., M.I. Mishchenko, A. Lacis, S. Zhang, J. Perlwitz, and S.M. Metzger, 2007. Do sulfate and nitrate coatings on mineral dust have important effects on radiative properties and climate modeling? *Journal of Geophysical Research,* **112**, D06307, doi:10.1029/2005JD006977.

Berntsen, T.K., J.S. Fuglestvedt , M.M. Joshi, K.P. Shine, N. Stuber, M. Ponater, R. Sausen, D.A. Hauglustaine, and L. Li, 2005: Response of climate to regional emissions of ozone precursors: sensitivities and warming potentials. *Tellus B,* **57(4)**, 283-304.

Boer, G. and B. Yu, 2003: Climate sensitivity and response. *Climate Dynamics,* **20(4)**, 415-429.

Bousquet, P., P. Ciais, J.B. Miller, E.J. Dlugokencky, D.A. Hauglustaine, C. Prigent, G.R. Van der Werf, P. Peylin, E.-G. Brunke, C. Carouge, R.L. Langenfelds, J. Lathière, F. Papa, M. Ramonet, M. Schmidt., L.P. Steele, S.C. Tyler, and J. White, 2006: Contribution of anthropogenic and natural sources to atmospheric methane variability. *Nature,* **443(7110)**, 439-442.

Butchart, N., A.A. Scaife, M. Bourqui, J. de Grandpré, S.H.E. Hare, J. Kettleborough, U. Langematz, E. Manzini, F. Sassi, K. Shibata, D. Shindell, and M. Sigmond, 2006: A multi-model study of climate change in the Brewer-Dobson circulation. *Climate Dynamics,* **27(7-8)**, 727-741.

Clarke, L., J. Edmonds, H. Jacoby, H. Pitcher, J. Reilly, and R. Richels, 2007: *Scenarios of Greenhouse Gas Emissions and Atmospheric Concentrations.* Sub-report 2.1A of Synthesis and Assessment Product 2.1 by the U.S. Climate Change Science Program and the Subcommittee on Global Change Research. Department of Energy, Office of Biological & Environmental Research, Washington, DC, 154 pp.

Collins, W.D., P.J. Rasch, B.E. Eaton, B. Khattatov, J.-F. Lamarque, and C.S. Zender, 2001: Simulating aerosols using a chemical transport model with assimilation of satellite aerosol retrievals: methodology for INDOEX. *Journal of Geophysical Research,* **106(D7)**, 7313-7336.

Collins, W.D., P.J. Rasch, B.A. Boville, J.J. Hack, J.R. McCaa, D.L. Williamson, B.P. Briegleb, C.M. Bitz, S.-J. Lin, and M. Zhang, 2006: The formulation and atmospheric simulation of the Community Atmosphere Model: CAM3. *Journal of Climate,* **19(5)**, 2144-2161.

Cox, S., W.-C. Wang, and S. Schwartz, 1995: Climate response to radiative forcings by sulfate aerosols and greenhouse gases. *Geophysical Research Letters,* **22(18)**, 2509-2512.

Delworth, T.L., A.J. Broccoli, A. Rosati, R.J. Stouffer, V. Balaji, J.A. Beesley, W.F. Cooke, K.W. Dixon, J. Dunne, K.A. Dunne, J.W. Durachta, K.L. Findell, P. Ginoux, A. Gnanadesikan, C.T. Gordon, S.M. Griffies, R. Gudgel, M.J. Harrison, I.M. Held, R.S. Hemler, L.W. Horowitz, S.A. Klein, T.R. Knutson, P.J. Kushner, A.R. Langenhorst, H.-C. Lee, S.-J. Lin, J. Lu, S.L. Malyshev, P.C.D. Milly, V. Ramaswamy, J. Russell, M.D. Schwarzkopf, E. Shevliakova, J.J. Sirutis, M.J. Spelman, W.F. Stern, M. Winton, A.T. Wittenberg, B. Wyman, F. Zeng, and R. Zhang, 2006: GFDL's CM2 global coupled climate models. Part I: Formulation and simulation characteristics. *Journal of Climate*, **19(5)**, 643-674.

Dentener, F.D., D.S. Stevenson, J. Cofala, R. Mechler, M. Amann, P. Bergamaschi, F. Raes, and R.G. Derwent, 2005: Tropospheric methane and ozone in the period 1990-2030: CTM calculations on the role of air pollutant and methane emissions controls. *Atmospheric Chemistry and Physics*, **5(7)**, 1731-1755.

Dlugokencky, E.J., S. Houweling, L. Bruhwiler, K.A. Masarie, P.M. Lang, J.B. Miller, and P.P. Tans, 2003: Atmospheric methane levels off: Temporary pause or a new steady-state? *Geophysical Research Letters*, **30(19)**, doi:10.1029/2003GL018126.

Fiore, A.M., L.W. Horowitz, E.J. Dlugokencky, and J.J. West, 2006: Impact of meteorology and emissions on methane trends, 1990-2004. *Geophysical Research Letters*, **33(12)**, L12809, doi:10.1029/2006GL026199.

Gauss, M., G. Myhre, G. Pitari, J.J. Prather, I.S.A. Isaksen, T.K. Berntsen, G.P. Brasseur, F.J. Dentener, R.G. Derwent, D.A. Hauglustaine, L.W. Horowitz, D.J. Jacob, M. Johnson, K.S. Law, L.J. Mickley, J.-F. Müller, P.-H. Plantevin, J.A. Pyle, H.L. Rogers, D.S. Stevenson, J.K. Sundet, M. van Weele, and O. Wild, 2003: Radiative forcing in the 21st century due to ozone changes in the troposphere and the lower stratosphere. *Journal of Geophysical Research*, **108(D9)**, 4292, doi:10.1029/2002JD002624.

Ginoux, P., L.W. Horowitz, V. Ramaswamy, I.V. Geogdzhayev, B.N. Holben, G. Stenchikov, and X. Tie, 2006: Evaluation of aerosol distribution and optical depth in the Geophysical Fluid Dynamics Laboratory coupled model CM2.1 for present climate. *Journal of Geophysical Research*, **111(D22)**, D22210, doi:10.1029/2005JD006707.

Guenther, A., C.N. Hewitt, D. Erickson, R. Fall, C. Geron, T. Graedel, P. Harley, L. Klinger, M. Lerdau, W.A. McKay, T. Pierce, B. Scholes, R. Steinbrecher, R. Tallamraju, J. Taylor and P. Zimmerman, 1995: A global model of natural volatile organic compound emissions *Journal of Geophysical Research*, **100(D5)**, 8873-8892.

Gustafson, W.I. and L. R. Leung, 2007: Regional downscaling for air quality assessment: A reasonable proposition? *Bulletin of the American Meteorological Society*, **88(8)**, 1215-1227.

Hansen, J.E., M. Sato, and R. Reudy, 1997: Radiative forcing and climate response. *Journal of Geophysical Research*, **102(D6)**, 6831-6864.

Hansen, J., M. Sato, R. Ruedy, L. Nazarenko, A. Lacis, G. A. Schmidt, G. Russell, I. Aleinov, M. Bauer, S. Bauer, N. Bell, B. Cairns, V. Canuto, M. Chandler, Y. Cheng, A. Del Genio, G. Faluvegi, E. Fleming, A. Friend, T. Hall, C. Jackman, M. Kelley, N. Kiang, D. Koch, J. Lean, J. Lerner, K. Lo, S. Menon, R. Miller, P. Minnis, T. Novakov, V. Oinas, Ja. Perlwitz, Ju. Perlwitz, D. Rind, A. Romanou, D. Shindell, P. Stone, S. Sun, N. Tausnev, D. Thresher, B. Wielicki, T. Wong, M. Yao, and S. Zhang, 2005: Efficacy of climate forcings. *Journal of Geophysical Research*, **110(D18)**, D18104, doi:10.1029/2005JD005776.

Haywood, J.M., R.J. Stouffer, R.T. Wetherald, S. Manabe, and V. Ramaswamy, 1997: Transient response of a coupled model to estimated changes in greenhouse gases and sulfate concentrations. *Geophysical Research Letters*, **24(11)**, 1335-1338.

Hegerl, G.C., F.W. Zwiers, P. Braconnot, N.P. Gillett, Y. Luo, J.A. Marengo Orsini, N. Nicholls, J.E. Penner, and P.A. Scott, 2007: Understanding and attributing climate change. In: *Climate Change 2007: The Physical Basis.* Contribution of Working Group I to the Fourth Assessment Report (AR4) of the Intergovernmental Panel on Climate Change [Solomon, S., D. Qin, M. Manning, Z. Chen, M. Marquis, K.B. Averyt, M. Tignor, and H.L. Miller (eds.)]. Cambridge University Press, Cambridge, UK, and New York, pp. 663-745.

Held, I.M. and B.J. Soden, 2006: Robust responses of the hydrological cycle to global warming. *Journal of Climate*, **19(21)**, 5686-5699.

Hogrefe, C., B. Lynn, K. Civerolo, J.-Y. Ku, J. Rosenthal, C. Rosenzweig, R. Goldberg, S. Gaffin, K. Knowlton, and P.L. Kinney, 2004: Simulating changes in regional air pollution over the eastern United States due to changes in global and regional climate and emissions. *Journal of Geophysical Research*, **109(D22)**, D22301, doi:10.1029/2004JD004690.

Horowitz, L.W., 2006: Past, present, and future concentrations of tropospheric ozone and aerosols: methodology, ozone evaluation, and sensitivity to aerosol wet removal. *Journal of Geophysical Research*, **111(22)**, D22211, doi:10.1029/2005JD006937.

Horowitz, L.W., S. Walters, D.L. Mauzerall, L.K. Emmons, P.J. Rasch, C. Granier, X. Tie, J.-F. Lamarque, M.G. Schultz, G.S. Tyndall, J.J. Orlando, and G.P. Brasseur, 2003: A global simulation of tropospheric ozone and related tracers: description and evaluation of MOZART, version 2. *Journal of Geophysical Research*, **108(D24)**, 4784, doi:10.1029/2002JD002853.

IPCC (Intergovernmental Panel on Climate Change), 2007: *Climate Change 2007: The Physical Science Basis.* Contribution of Working Group I to the Fourth Assessment Report (AR4) of the Intergovernmental Panel on Climate Change [Solomon, S., D. Qin, M. Manning, Z. Chen, M. Marquis, K.B. Averyt, M. Tignor, and H.L. Miller (eds.)]. Cambridge University Press, Cambridge, UK and New York, 996 pp.

Kaufman, Y.J., I. Koren, L.A. Remer, D. Rosenfeld, and Y. Rudich, 2005: The effect of smoke, dust, and pollution aerosol on shallow cloud development over the Atlantic Ocean. *Proceedings of the National Academy of Sciences*, **102(32)**, 11207-11212.

Kiehl, J.T., T.L. Schneider, R.W. Portmann, and S. Solomon, 1999: Climate forcing due to tropospheric and stratospheric ozone. *Journal of Geophysical Research*, **104(D24)**, 31239-31254.

Kinne, S., M. Schulz, C. Textor, S. Guibert, Y. Balkanski, S.E.. Bauer, T. Berntsen, T.F. Berglen, O. Boucher, M. Chin, W. Collins, F. Dentener, T. Diehl, R. Easter, J. Feichter, D. Fillmore, S. Ghan, P. Ginoux, S. Gong, A. Grini, J. Hendricks, M. Herzog, L. Horowitz, I. Isaksen, T. Iversen, A. Kirkevåg, S. Kloster, D. Koch, J.E. Kristjansson, M. Krol, A. Lauer, J.F. Lamarque, G. Lesins, X. Liu, U. Lohmann, V. Montanaro, G. Myhre, J. Penner, G. Pitari, S. Reddy, O. Seland, P. Stier, T. Takemura, and X. Tie, 2006: An AeroCom initial assessment – optical properties in aerosol component modules of global models. *Atmospheric Chemistry and Physics*, **6(7)**, 1815-1834.

Knutson, T.R., T.L. Delworth, K.W. Dixon, I.M. Held, J. Lu, V. Ramaswamy, M.D. Schwarzkopf, G. Stenchikov, and R.J. Stouffer, 2006: Assessment of twentieth-century regional surface temperature trends using the GFDL CM2 coupled models. *Journal of Climate*, **10(9)**, 1624-1651.

Koch, D., G. Schmidt, and C. Field, 2006: Sulfur, sea salt and radionuclide aerosols in GISS ModelE. *Journal of Geophysical Research*, **111(D6)**, D06206, doi:10.1029/2004JD005550.

Koch, D., T. Bond, D. Streets, N. Bell, and G.R. van der Werf, 2007: Global impacts of aerosols from particular source regions and sectors. *Journal of Geophysical Research*, **112(D2)**, D02205, doi:10.1029/2005JD007024.

Lamarque, J.-F., P. Hess, L. Emmons, L. Buja, W.M. Washington, and C. Granier, 2005a: Tropospheric ozone evolution between 1890 and 1990. *Journal of Geophysical Research*, **110(D8)**, D08304, doi:10.1029/2004JD005537.

Lamarque, J.-F., J.T. Kiehl, P.G. Hess, W.D. Collins, L.K. Emmons, P. Ginoux, C. Luo, and X.X. Tie, 2005b: Response of a coupled chemistry-climate model to changes in aerosol emissions: global impact on the hydrological cycle and the tropospheric burdens of OH, ozone and NOx. *Geophysical Research Letters*, **32(16)**, L16809, doi:10.1029/2005GL023419.

Leung, L.R. and W.I. Gustafson, 2005: Potential regional climate change and implications to US air quality. *Geophysical Research Letters*, **32(16)**, L16711, doi:10.1029/2005GL022911.

Levy II, H., M.D. Schwarzkopf, L. Horowitz, V. Ramaswamy, and K.L. Findell, 2008: Strong sensitivity of late 21st century climate to projected changes in short-lived air pollutants. *Journal of Geophysical Research*, **113**, DO6102, doi:10.1029/2007JD009176.

Liang, X.Z., L. Li, K.E. Kunkel, M. Ting, and J.X.L. Wang, 2004: Regional climate model simulation of U.S. precipitation during 1982–2002. Part I: Annual cycle. *Journal of Climate*, **17(18)**, 3510-3529.

Liang, X.-Z., J. Pan, J. Zhu, K.E. Kunkel, J.X.L. Wang, and A. Dai, 2006: Regional climate model downscaling of the U.S. summer climate and future change. *Journal of Geophysical Research*, **111(D10)**, D10108, doi:10.1029/2005JD006685.

Lohmann, U. and G. Lesins, 2002: Stronger constraints on the anthropogenic indirect aerosol effect. *Science*, **298**, 1012-1015.

Lohmann, U., I. Koren, and Y.J. Kaufman, 2006: Disentangling the role of microphysical and dynamical effects in determining cloud properties over the Atlantic. *Geophysical Research Letters*, **33**, L09802, doi:10.1029/2005GL024625.

Mahowald, N. M. and C. Luo, 2003: A less dusty future? *Geophysical Research Letters*, **30**, 1903, doi:10.1029/2003GL017880.

Menon, S.A., D. Del Genio, D. Koch, and G. Tselioudis, 2002: GCM simulations of the aerosol indirect effect: sensitivity to cloud parameterization and aerosol burden. *Journal of the Atmospheric Sciences*, **59(3)**, 692-713.

Miller, R.L., R.V. Cakmur, J. Perlwitz, I.V. Geogdzhayev, P. Ginoux, D. Koch, K.E. Kohfeld, C. Prigent, R. Ruedy, G.A. Schmidt, and I. Tegen, 2006a: Mineral dust aerosols in the NASA Goddard Institute for Space Studies ModelE AGCM. *Journal of Geophysical Research*, **111(D18)**, D0208, doi:10.1029/2005JD005796.

Miller, R.L., G.A. Schmidt, and D.T. Shindell, 2006b: Forced variations of annular modes in the 20th century Intergovernmental Panel on Climate Change Fourth Assessment Report models. *Journal of Geophysical Research*, **111**, D18101, doi:10.1029/2005JD006323.

Mitchell, J.F.B., R.A. Davis, W.J. Ingram, and C.A. Senior, 1995: On surface temperature, greenhouse gases, and aerosols: models and observations. *Journal of Climate*, **8(10)**, 2364-2386.

Nakićenović, N. and R. Swart (eds.), 2000: *Special Report on Emissions Scenarios*. A special report of Working Group III of the Intergovernmental Panel on Climate Change. Cambridge University Press, Cambridge, UK, and New York, 599 pp.

Nolte, C.G., A.B. Gilliland, C. Hogrefe, and L.J. Mickley, 2008: Linking global to regional models to assess future climate impacts on surface ozone levels in the United States. *Journal of Geophysical Research*, **113**, D14307, doi:10.1029/2007JD008497.

Olivier, J.G.J. and J.J.M. Berdowski, 2001: Global emissions sources and sinks. In: *The Climate System* [Berdowski, J., R. Guicherit and B.-J. Heij (eds.)]. A.A. Balkema Publishers/ Swets & Zeitlinger Publishers, Lisse, The Netherlands, pp. 33-78.

Pincus, R. and M. Baker, 1994: Precipitation, solar absorption, and albedo susceptibility in marine boundary layer clouds. *Nature*, **372(6503)**, 250-252.

Ramaswamy, V. and C.-T. Chen, 1997: Linear additivity of climate response for combined albedo and greenhouse perturbations. *Geophysical Research Letters*, **24(5)**, 567-570.

Ramaswamy, V., O. Boucher, J. Haigh, D. Hauglustaine, J. Haywood, G. Myhre, T. Nakajima, G.Y. Shi, and S. Solomon, 2001: Radiative forcing of climate change. In: *Climate Change 2001: The Scientific Basis*. Contribution of Working Group I to the Third Assessment Report of the Intergovernmental Panel on Climate Change [Houghton, J. T., Y. Ding, D.J. Griggs, M. Noguer, P.J. van der Linden, X. Dai, K. Maskell, and C.A. Johnson (eds.)]. Cambridge University Press, Cambridge, UK, and New York, pp. 349-416.

Schmidt, G.A., R. Ruedy, J.E. Hansen, I. Aleinov, N. Bell, M. Bauer, S. Bauer, B. Cairns, V. Canuto, Y. Cheng, A. Del Genio, G. Faluvegi, A.D. Friend, T.M. Hall, Y. Hu, M. Kelley, N.Y. Kiang, D. Koch, A.A. Lacis, J. Lerner, K.K. Lo, R.L. Miller, L. Nazarenko, V. Oinas, J. Perlwitz, J. Perlwitz, D. Rind, A. Romanou, G.L. Russell, M. Sato, D.T. Shindell, P.H. Stone, S. Sun, N. Tausnev, D. Thresher, and M.-S. Yao, 2006: Present day atmospheric simulations using GISS ModelE: comparison to in-situ, satellite and reanalysis data. *Journal of Climate*, **19(2)**, 153-192.

Schulz, M., C. Textor, S. Kinne, Y. Balkanski, S. Bauer, T. Berntsen, T. Berglen, O. Boucher, F. Dentener, S. Guibert, I.S.A.

Isaksen, T. Iversen, D. Koch, A. Kirkevåg, X. Liu, V. Montanaro, G. Myhre, J.E. Penner, G. Pitari, S. Reddy, Ø. Seland, P. Stier, and T. Takemura, 2006: Radiative forcing by aerosols as derived from the AeroCom present-day and pre-industrial simulations. *Atmospheric Chemistry and Physics*, **6(12)**, 5225-5246.

Shindell, D.T., G. Faluvegi, and N. Bell, 2003: Preindustrial-to-present-day radiative forcing by tropospheric ozone from improved simulations with the GISS chemistry-climate GCM. *Atmospheric and Chemical Physics*, **3(5)**, 1675-1702.

Shindell, D.T., B.P. Walter, and G. Faluvegi, 2004: Impacts of climate change on methane emissions from wetlands. *Geophysical Research Letters*, **31(21)**, L21202, doi:10.1029/2004GL021009.

Shindell, D.T., G. Faluvegi, A. Lacis, J.E. Hansen, R. Ruedy, and E. Aguilar, 2006a: The role of tropospheric ozone increases in 20th century climate change. *Journal of Geophysical Research*, **111(D8)**, D08302, , doi:10.1029/2005JD006348.

Shindell, D.T., G. Faluvegi, N. Unger, E. Aguilar, G.A. Schmidt, D. Koch, S.E. Bauer, and R.L. Miller, 2006b: Simulations of preindustrial, present-day, and 2100 conditions in the NASA GISS composition and climate model G-PUCCINI. *Atmospheric Chemistry and Physics*, **6**, 4427-4459.

Shindell, D.T., G. Faluvegi, S.E. Bauer, D.M. Koch, N. Unger, S. Menon, R.L. Miller, G.A. Schmidt, and D.G. Streets, 2007: Climate response to projected changes in short-lived species under an A1B scenario from 2000-2050 in the GISS climate model. *Journal of Geophysical Research*, **112(D20)**, D20103, doi:10.1029/2007JD008753.

Shindell, D.T., H. Levy II, M.D. Schwarzkopf, L.W. Horowitz, J.-F. Lamarque, and G. Faluvegi, 2008: Multi-model projections of climate change from short-lived emissions due to human activities. *Journal of Geophysical Research*, **113**, D11109, doi:10.1029/2007JD009152.

Stevenson, D.S., F.J. Dentener, M. G. Schultz, K. Ellingsen, T.P.C. van Noije, O. Wild, G. Zeng, M. Amann, C.S. Atherton, N. Bell, D.J. Bergmann, I. Bey, T. Butler, J. Cofala, W.J. Collins, R.G. Derwent, R.M. Doherty, J. Drevet, H.J. Eskes, A.M. Fiore, M. Gauss, D.A. Hauglustaine, L.W. Horowitz, I.S.A. Isaksen, M.C. Krol, J.-F. Lamarque, M.G. Lawrence, V. Montanaro, J.-F. Müller, G. Pitari, M.J. Prather, J.A. Pyle, S. Rast, J.M. Rodriguez, M.G. Sanderson, N.H. Savage, D.T. Shindell, S.E. Strahan, K. Sudo, and S. Szopa, 2006: Multi-model ensemble simulations of present-day and near-future tropospheric ozone. *Journal of Geophysical Research*, **111(D8)**, D08301, doi:10.1029/2005JD006338.

Stewart, R.W., S. Hameed, and J.P. Pinto, 1977: Photochemistry of the tropospheric ozone. *Journal of Geophysical Research*, **82(21)**, 3134-3140.

Stouffer, R.J., 2004: Time scales of climate response. *Journal of Climate*, **17(1)**, 209-217.

Stouffer, R.J., T.L. Delworth, K.W. Dixon, R. Gudgel, I. Held, R. Hemler, T. Knutson, M.D. Schwarzkopf, M.J. Spelman, M.W. Winton, A.J. Broccoli, H-C. Lee, F. Zeng, and B. Soden, 2006: GFDL's CM2 global coupled climate models. Part IV: Idealized climate response. *Journal of Climate*, **19(5)**, 723-740.

Streets, D., T.C. Bond, T. Lee, and C. Jang, 2004: On the future of carbonaceous aerosol emissions. *Journal of Geophysical Research*, **109(D24)**, doi:10.1029/2004JD004902.

Twomey, S., 1974: Pollution and the planetary albedo. *Atmospheric Environment*, **8(12)**, 1251-1256.

Unger, N., D.T. Shindell, D.M. Koch, and D.G. Streets, 2008: Air pollution radiative forcing from specific emissions sectors at 2030. *Journal of Geophysical Research*, **113**, D02306, doi:10.1029/2007JD008683.

Van der Werf, G.R., J.T. Randerson, G.J. Collatz, and L. Giglio, 2003: Carbon emissions from fires in tropical and subtropical ecosystems. *Global Change Biology*, **9(4)**, 547-562.

Woodward, S., D.L. Roberts, and R.A. Betts, 2005: A simulation of the effect of climate change-induced desertification on mineral dust aerosol. *Geophysical Research Letters*, **32(18)**, L18810, doi:10.1029/2005GL023482.

CHAPTER 4 REFERENCES

Bauer, S.E. and D. Koch, 2005: Impact of heterogeneous sulfate formation at mineral dust surfaces on aerosol loads and radiative forcing in the Goddard Institute for Space Studies general circulation model. *Journal of Geophysical Research*, **110(D17)**, D17202, doi:10.1029/2005JD005870.

Bauer, S.E., D. Koch, N. Unger, S.M. Metzger, D.T. Shindell, and D. Streets, 2007: Nitrate aerosols today and in 2030: importance relative to other aerosol species and tropospheric ozone. *Atmospheric Chemistry and Physics*, **7(19)**, 5043-5059.

Berntsen, T.K., J.S. Fuglestvedt, M.M. Joshi, K.P. Shine, N. Stuber, M. Ponater, R. Sausen, D.A. Hauglustaine, and L. Li, 2005: Response of climate to regional emissions of ozone precursors: sensitivities and warming potentials. *Tellus B*, **57(4)**, 283-304.

Clarke, L., J. Edmonds, H. Jacoby, H. Pitcher, J. Reilly, and R. Richels, 2007: *Scenarios of Greenhouse Gas Emissions and Atmospheric Concentrations*. Sub-report 2.1A of Synthesis and Assessment Product 2.1 by the U.S. Climate Change Science Program and the Subcommittee on Global Change Research. Department of Energy, Office of Biological & Environmental Research, Washington, DC, 154 pp.

Fan, S-M., L.W. Horowitz, H. Levy II, and W.J. Moxim, 2004: Impact of air pollution on wet deposition of mineral dust aerosols. *Geophysical Research Letters*, **31(2)**, L02104, doi:10.1029/2003GL018501.

Fiore, A.M., D.J. Jacob, B.D. Field, D.G. Streets, S.D. Fernandes, and C. Jang, 2002: Linking ozone pollution and climate change: The case for controlling methane. *Geophysical Research Letters*, **29(19)**, 1919, doi:10.1029/2002GL015601.

Hegerl, G.C., F.W. Zwiers, P. Braconnot, N.P. Gillett, Y. Luo, J.A. Marengo Orsini, N. Nicholls, J.E. Penner, and P.A. Scott, 2007: Understanding and attributing climate change. In: *Climate Change 2007: The Physical Science Basis*. Contribution of Working Group I to the Fourth Assessment Report (AR4) of the Intergovernmental Panel on Climate Change [Solomon, S., D. Qin, M. Manning, Z. Chen, M. Marquis, K.B.Averyt, M. Tignor and H.L. Miller (eds.)]. Cambridge University Press, Cambridge, UK and New York, pp. 663-745.

IPCC (Intergovernmental Panel on Climate Change) 2007: *Climate Change 2007: The Physical Science Basis*. Contribution of Working Group I to the Fourth Assessment Report (AR4) of the Intergovernmental Panel on Climate Change [Solomon, S., D. Qin, M. Manning, Z. Chen, M. Marquis, K.B. Averyt, M.Tignor, and H.L. Miller (eds.)]. Cambridge University Press, Cambridge, UK, and New York, 996 pp.

Mahowald, N.M. and C. Luo, 2003: A less dusty future? *Geophysical Research Letters*, **30(17)**, doi:10.1029/2003GL017880.

Shindell, D.T., G. Faluvegi, N. Bell, and G.A. Schmidt, 2005: An emissions-based view of climate forcing by methane and tropospheric ozone. *Geophysical Research Letters*, **32**, L04803, doi:10.1029/2004GL021900.

Streets, D., T.C. Bond, T. Lee, and C. Jang, 2004: On the future of carbonaceous aerosol emissions. *Journal of Geophysical Research*, **109(D24),** D24212, doi:10.1029/2004JD004902.

West, J.J. and A.M. Fiore, 2005: Management of tropospheric ozone by reducing methane emissions. *Environmental Science & Technology*, **39**, 4685-4691.

West, J.J., A.M. Fiore, L.W. Horowitz, and D.L. Mauzerall, 2006: Global health benefits of mitigating ozone pollution with meth-

ane emission controls. *Proceedings of the National Academy of Sciences*, **103(11)**, 3988-3993.

West, J.J., A.M. Fiore, V. Naik, L.W. Horowitz, M.D. Schwarzkopf, and D.L. Mauzerall, 2007: Ozone air quality and radiative forcing consequences of changes in ozone precursor emissions. *Geophysical Research Letters*, **34**, L06806, doi:10.1029/2006GL029173.

Woodward, S., D.L. Roberts, and R.A. Betts, 2005: A simulation of the effect of climate change-induced desertification on mineral dust aerosol. *Geophysical Research Letters*, **32(18)**, L18810, doi:10.1029/2005GL023482.

APPENDIX A REFERENCES

Clarke, L., J. Edmonds, H. Jacoby, H. Pitcher, J. Reilly, and R. Richels, 2007: *Scenarios of Greenhouse Gas Emissions and Atmospheric Concentrations*. Sub-report 2.1A of Synthesis and Assessment Product 2.1 by the U.S. Climate Change Science Program and the Subcommittee on Global Change Research. Department of Energy, Office of Biological & Environmental Research, Washington, DC, 154 pp.

CCSP (Climate Change Science Program), 2008: *Weather and Climate Extremes in a Changing Climate: Regions of Focus: North America, Hawaii, Caribbean, and U.S. Pacific Islands.* [Karl, T.R., G.A. Meehl, C.D. Miller, S.J. Hassol, A.M. Waple, and W.L. Murray (eds.)]. Synthesis and Assessment Product 3.3. U.S. Climate Change Science Program, Washington, DC, 164 pp.

Dai, A., G.A. Meehl, W.M. Washington, T.M.L. Wigley, and J.M. Arblaster, 2001a: Ensemble simulation of twenty-first century climate changes: business as usual vs. CO_2 stabilization. *Bulletin of the American Meteorological Society*, **82(11)**, 2377-2388.

Dai, A., T.M.L. Wigley, G.A. Meehl, and W.M. Washington, 2001b: Effects of stabilizing atmospheric CO_2 on global climate in the next two centuries. *Geophysical Research Letters*, **28(23)**, 4511-4514.

Emanuel, K., R. Sundararajan, and J. Williams, 2008: Hurricanes and global warming: results from downscaling IPCC AR4 simulations. *Bulletin of American Meteorological Society*, **89(3)**, 347-367.

Knutson, T.R., J.J. Sirutis, S.T. Garner, G.A. Vecchi, and I.M. Held, 2008: Simulated reduction in Atlantic hurricane frequency under twenty-first century warming conditions. *Nature Geoscience*, **1(6)**, 359-364.

Meehl, G.A., T.F. Stocker, W.D. Collins, P. Friedlingstein, A.T. Gaye, J.M. Gregory, A. Kitoh, R. Knutti, J.M. Murphy, A. Noda, S.C.B. Raper, I.G. Watterson, A.J. Weaver and Z.-C. Zhao, 2007: Global climate projections. In: *Climate Change 2007: The Physical Basis*. Contribution of Working Group I to the Fourth Assessment Report (AR4) of the Intergovernmental Panel on Climate Change [Solomon, S., D. Qin, M. Manning, Z. Chen, M. Marquis, K.B. Averyt, M. Tignor, and H.L. Miller (eds.)]. Cambridge University Press, Cambridge, UK, and New York, pp. 747-845.

Wigley, T.M.L., R. Richels, and J.A. Edmonds, 2007: Overshoot pathways to CO_2 stabilization in a multi-gas context. In: *Human Induced Climate Change: An Interdisciplinary Assessment* [Schlesinger, M., F.C. de la Chesnaye, H. Kheshgi, C.D. Kolstad, J. Reilly, J.B. Smith, and T. Wilson (eds.)]. Cambridge University Press, Cambridge, UK, pp. 84-92.

APPENDIX B REFERENCES

Church, J.A. and J.M. Gregory, 2001: Changes in sea level. *In: Climate Change 2001: The Scientific Basis*. Contribution of Working Group I to the Third Assessment Report of the Intergovernmental Panel on Climate Change [Houghton, J.T., Y. Ding, D.J. Griggs, M. Noguer, P.J. van der Linden, X. Dai, K. Maskell, and C.A. Johnson (eds.)]. Cambridge University Press, Cambridge, UK, and New York, pp. 639-693.

Cubasch, U. and G.A. Meehl, 2001: Projections for future climate change. In: *Climate Change 2001: The Scientific Basis*. Contribution of Working Group I to the Third Assessment Report of the Intergovernmental Panel on Climate Change [Houghton, J.T., Y. Ding, D.J. Griggs, M. Noguer, P.J. van der Linden, X. Dai, K. Maskell, and C.A. Johnson (eds.)]. Cambridge University Press, Cambridge, UK, and New York, pp. 525-582.

Harvey, L.D.D., J. Gregory, M. Hoffert, A. Jain, M. Lal, R. Leemans, S.B.C. Raper, T.M.L. Wigley, and J. de Wolde, 1997: *An Introduction to Simple Climate Models Used in the IPCC Second Assessment Report*. IPCC technical paper 2 [Houghton, J.T., L.G. Meira Filho, D.J. Griggs, and M. Noguer (eds.)]. Intergovernmental Panel on Climate Change, Geneva, Switzerland, 50 pp.

Hoffert, M.L., A.J. Callegari, and C.-T. Hsieh, 1980: The role of deep sea heat storage in the secular response to climate forcing. *Journal of Geophysical Research*, **85(C11)**, 6667-6679.

IPCC (Intergovernmental Panel on Climate Change), 2001: *Climate Change 2001: The Scientific Basis*. Contribution of Working Group I to the Third Assessment Report of the Intergovernmental Panel on Climate Change [Houghton, J.T., Y. Ding, D.J. Griggs, M. Noguer, P.J. van der Linden, X. Dai,

K. Maskell, and C.A. Johnson (eds.)]. Cambridge University Press, Cambridge, UK, and New York, 881 pp.

Joos, F., I.C. Prentice, S. Sitch, R. Meyer, G. Hooss, G.-K. Plattner, S. Gerber, and K. Hasselmann, 2001: Global warming feedbacks on terrestrial carbon uptake under the Intergovernmental Panel on Climate Change (IPCC) emissions scenarios. *Global Biogeochemical Cycles* **15(4)**, 891-908, doi:10.1029/2000GB001375.

Kheshgi, H.S. and A.K. Jain, 2003: Projecting future climate change: implications of carbon cycle model intercomparisons. *Global Biogeochemical Cycles* **17(2)**, 1047, doi:10.1029/2001GB001842.

Meehl, G.A., T.F. Stocker, W.D. Collins, P. Friedlingstein, A.T. Gaye, J.M. Gregory, A. Kitoh, R. Knutti, J.M. Murphy, A. Noda, S.C.B. Raper, I.G. Watterson, A.J. Weaver and Z.-C. Zhao, 2007: Global climate projections. In: *Climate Change 2007: The Physical Science Basis.* Contribution of Working Group I to the Fourth Assessment Report (AR4) of the Intergovernmental Panel on Climate Change [Solomon, S., D. Qin, M. Manning, Z. Chen, M. Marquis, K.B.Averyt, M. Tignor and H.L. Miller (eds.)]. Cambridge University Press, Cambridge, United Kingdom and New York, pp. 747-845.

Prather, M. and D. Ehhalt, 2001: Atmospheric chemistry and greenhouse gases. In: *Climate Change 2001: The Scientific Basis.* Contribution of Working Group I to the Third Assessment Report of the Intergovernmental Panel on Climate Change [Houghton, J. T., Y. Ding, D.J. Griggs, M. Noguer, P.J. van der Linden, X. Dai, K. Maskell, and C.A. Johnson (eds.)]. Cambridge University Press, Cambridge, UK, and New York, pp. 239-287.

Raper, S.C.B., J.M. Gregory, and T.J. Osborn, 2001: Use of an upwelling-diffusion energy balance climate model to simulate and diagnose A/OGCM results. *Climate Dynamics,* **17(8)**, 601-613.

Wigley, T.M.L., 1989: Possible climatic change due to SO_2-derived cloud condensation nuclei. *Nature,* **339(6223)**, 365-367.

Wigley, T.M.L., 1991a: A simple inverse carbon cycle model. *Global Biogeochemical Cycles,* **5(4)**, 373-382.

Wigley, T.M.L., 1991b: Could reducing fossil-fuel emissions cause global warming? *Nature,* **349(6309)**, 503-506.

Wigley, T.M.L., 1993: Balancing the carbon budget. Implications for projections of future carbon dioxide concentration changes. *Tellus B,* **45(5)**, 409-425.

Wigley, T.M.L., 2000: Stabilization of CO_2 concentration levels. In: *The Carbon Cycle* [Wigley, T.J.L. and D.S. Schimel (ed.)]. Cambridge University Press, Cambridge, UK, and New York, pp. 258-276.

Wigley, T.M.L. and S.C.B. Raper, 2001: Interpretation of high projections for global-mean warming. *Science,* **293(5529)**, 451-454.

Wigley, T.M.L. and S.C.B. Raper, 2005: Extended scenarios for glacier melt due to anthropogenic forcing. *Geophysical Research Letters,* **32(5)**, L05704, doi:10.1029/2004GL021238.

Wigley, T.M.L., S.J. Smith, and M.J. Prather, 2002: Radiative forcing due to reactive gas emissions. *Journal of Climate,* **15(18)**, 2690-2696.

Wigley, T.M.L., R. Richels, and J.A. Edmonds, 2007: Overshoot pathways to CO_2 stabilization in a multi-gas context. In: *Human Induced Climate Change: An Interdisciplinary Assessment* [Schlesinger, M., F.C. de la Chesnaye, H. Kheshgi, C.D. Kolstad, J. Reilly, J.B. Smith and T. Wilson (eds.)]. Cambridge University Press, Cambridge, UK, pp. 84-92.

APPENDIX C REFERENCES

Andreae, M.O. and P. Merlet, 2001: Emission of trace gases and aerosols from biomass burning. *Global Biogeochemical Cycles,* **15(4)**, 955-966.

Bauer, S.E., M.I. Mishchenko, A. Lacis, S. Zhang, J. Perlwitz, and S.M. Metzger, 2007: Do sulfate and nitrate coatings on mineral dust have important effects on radiative properties and climate modeling? *Journal of Geophysical Research,* **112(D6)**, D06307, doi: 10.1029/2005JD006977.

Bian, H. and M. Prather, 2002: Fast-J2: accurate simulations of photolysis in global climate models. *Journal of Atmospheric Chemistry,* **41(3)**, 281-296.

Brasseur, G.P., D.A. Hauglustaine, S. Walters, P.J. Rasch, J.-F. Müller, C. Granier, and X. Tie, 1998: MOZART, a global chemical transport model for ozone and related chemical tracers: 1. Model description. *Journal of Geophysical Research,* **103(D21)**, 28265-28289.

Chin, M., D.J. Jacob, G.M. Gardner, M.S. Forman-Fowler, P.A. Spiro, and D.L. Savoie, 1996: A global three-dimensional model of tropospheric sulfate. *Journal of Geophysical Research,* **101(D13)**, 18667-18690.

Chung, S.H. and J.H. Seinfeld, 2002: Global distribution and climate forcing of carbonaceous aerosols. *Journal of Geophysical Research*, **107(D19)**, doi: 10.1029/2001JD001397.

Collins, W.D., P.J. Rasch, B.E. Eaton, B. Khattatov, J.-F. Lamarque, and C.S. Zender, 2001: Simulating aerosols using a chemical transport model with assimilation of satellite aerosol retrievals: methodology for INDOEX. *Journal of Geophysical Research*, **106(D7)**, 7313-7336.

Collins, W.D., P.J. Rasch, B.A. Boville, J.J. Hack, J.R. McCaa, D.L. Williamson, B.P. Briegleb, C.M. Bitz, J.-J. Lin, and M. Zhang, 2006: The formulation and atmospheric simulation of the Community Atmosphere Model: CAM3. *Journal of Climate*, **19(11)**, 2144-2161.

Del Genio, A.D. and M.-S. Yao, 1993: Efficient cumulus parameterization for long-term climate studies: the GISS scheme. In: *The Representation of Cumulus Convection in Numerical Models* [Emanuel, K.A. and D.A. Raymond (eds.)]. AMS meteorological monograph v.24, no.46. American Meteorological Society, Boston, pp. 181-184.

Dentener, F.D., D.S. Stevenson, J. Cofala, R. Mechler, M. Amann, P. Bergamaschi, F. Raes, and R.G. Derwent, 2005:Tropospheric methane and ozone in the period 1990-2030: CTM calculations on the role of air pollutant and methane emissions controls. *Atmospheric Chemistry and Physics*, **5(7)**, 1731-1755.

Gery, M.W., G.Z. Whitten, J.P. Killus, and M.C. Dodge, 1989:A photochemical kinetics mechanism for urban and regional scale computer modeling. *Journal of Geophysical Research*, **94(D10)**, 925-956.

Ginoux, P., L.W. Horowitz, V. Ramaswamy, I.V. Geogdzhayev, B.N. Holben, G. Stenchikov, and X. Tie, 2006: Evaluation of aerosol distribution and optical depth in the Geophysical Fluid Dynamics Laboratory coupled model CM2.1 for present climate. *Journal of Geophysical Research*, **111(D22)**, D22210, doi:10.1029/2005JD006707.

Giorgi, F. and W.L. Chameides, 1985: The rainout parameterization in a photochemical model. *Journal of Geophysical Research*, **90(D5)**, 7872-7880.

Hack, J.J., 1994: Parameterization of moist convection in the NCAR community climate model (CCM2). *Journal of Geophysical Research*, **99(D3)**, 5551-5568.

Hanisch, F. and J.N. Crowley, 2001: The heterogeneous reactivity of gaseous nitric acid on authentic mineral dust samples, and on individual mineral and clay mineral components. *Journal of Physical Chemistry A*, **105 (13)**, 3096-3106.

Holtslag, A. and B. Boville, 1993: Local versus nonlocal boundary-layer diffusion in a global climate model. *Journal of Climate*, **6(10)**, 1825-1842.

Horowitz, L.W., 2006: Past, present, and future concentrations of tropospheric ozone and aerosols: Methodology, ozone evaluation, and sensitivity to aerosol wet removal. *Journal of Geophysical Research*, **111(22)**, D22211, doi:10.1029/2005JD006937.

Horowitz, L.W., S. Walters, D.L. Mauzerall, L.K. Emmons, P.J. Rasch, C. Granier, X. Tie, J.-F. Lamarque, M.G. Schultz, G.S. Tyndall, J.J. Orlando, and G.P. Brasseur, 2003: A global simulation of tropospheric ozone and related tracers: Description and evaluation of MOZART, version 2. *Journal of Geophysical Research*, **108(D24)**, 4784, doi:10.1029/2002JD002853.

Houweling, S., F. Dentener, and J. Lelieveld, 1998: The impact of non-methane hydrocarbon compounds on tropospheric photochemistry. *Journal of Geophysical Research*, **103,(D9)**, 1 0673-10696.

Kiehl, J.T., J.J. Hack, G.B. Bonan, B.A. Boville, D.L. Williamson, and P.J. Rasch, 1998: The National Center for Atmospheric Research Community Climate Model: CCM3. *Journal of Climate*, **11(6)**, 1131-1149.

Kiehl, J.T., T.L. Schneider, R.W. Portmann, and S. Solomon, 1999: Climate forcing due to tropospheric and stratospheric ozone. *Journal of Geophysical Research*, **104(D24)**, 31239-31254.

Koch, D., G. Schmidt, and C. Field, 2006: Sulfur, sea salt and radionuclide aerosols in GISS ModelE. *Journal of Geophysical Research*, **111(D6)**, D06206, doi:10.1029/2004JD005550.

Koch, D., T. Bond, D. Streets, N. Bell, and G.R. van der Werf, 2007: Global impacts of aerosols from particular source regions and sectors. *Journal of Geophysical Research*, **112(D2)**, D02205, doi:10.1029/2005JD007024.

Lamarque, J.-F., P. Hess, L. Emmons, L. Buja, W. M. Washington, and C. Granier, 2005: Tropospheric ozone evolution between 1890 and 1990. *Journal of Geophysical Research*, **110(D8)**, D08304, doi:10.1029/2004JD005537.

Lin, S.-J. and R.B. Rood, 1996: Multidimensional flux-form semi-Lagrangian transport schemes. *Monthly Weather Review*, **124(9)**, 2046-2070.

Madronich, S. and S. Flocke, 1998: The role of solar radiation in atmospheric chemistry. In: *Environmental Photochemistry* [Boule, P. (ed.)]. Handbook of Environmental Chemistry, v.2, Pt.L. Springer-Verlag, Berlin and New York, pp. 1-26.

Metzger, S., F. Dentener, S. Pandis, and J. Lelieveld, 2002: Gas/aerosol partitioning: 1. A computationally efficient model. *Journal of Geophysical Research*, **107(D16)**, doi:10.1029/2001jd001102.

Miller, R.L., R.V. Cakmur, J. Perlwitz, I.V. Geogdzhayev, P. Ginoux, D. Koch, K.E. Kohfeld, C. Prigent, R. Ruedy, G.A. Schmidt, and I. Tegen, 2006: Mineral dust aerosols in the NASA Goddard Institute for Space Studies ModelE AGCM. *Journal of Geophysical Research*, **111(D6)**, D0208, doi:10.1029/2005JD005796.

Nicolet, M., 1984: On the photodissociation of water vapour in the mesosphere. *Planetary and Space Science*, **32(7)**, 871-880.

Nicolet, M. and S. Cieslik, 1980: The photodissociation of nitric oxide in the mesosphere and stratosphere. *Planetary and Space Science*, **28(1)**, 105-115.

Olivier, J.G.J. and J.J.M. Berdowski, 2001: Global emissions sources and sinks. In: *The Climate System* [Berdowski, J., R. Guicherit and B.-J. Heij (eds.)]. A.A. Balkema Publishers/ Swets & Zeitlinger Publishers, Lisse, The Netherlands, pp. 33-78.

Sander, S.P., R.R. Friedl, W.B. DeMore, A.R. Ravishankara, D.M. Golden, C.E. Kolb, M.J. Kurylo, R.F. Hampson, R.E. Huie, M.J. Molina, and G.K. Moortgat, 2000: *Chemical Kinetics and Photochemical Data for Use in Stratospheric Modeling, Evaluation Number 13*. JPL publication 00-003. NASA Jet Propulsion Laboratory, Pasadena, CA, 74 pp. Available at http://jpldataeval.jpl.nasa.gov/previous_evaluations.html.

Schmidt, G.A., R. Ruedy, J.E. Hansen, I. Aleinov, N. Bell, M. Bauer, S. Bauer, B. Cairns, V. Canuto, Y. Cheng, A. Del Genio, G. Faluvegi, A.D. Friend, T.M. Hall, Y. Hu, M. Kelley, N.Y. Kiang, D. Koch, A.A. Lacis, J. Lerner, K.K. Lo, R.L. Miller, L. Nazarenko, V. Oinas, J. Perlwitz, J. Perlwitz, D. Rind, A. Romanou, G.L. Russell, M. Sato, D.T. Shindell, P.H. Stone, S. Sun, N. Tausnev, D. Thresher, and M.-S. Yao 2006: Present day atmospheric simulations using GISS ModelE: comparison to in-situ, satellite and reanalysis data. *Journal of Climate*, **19(2)**, 153-192.

Shindell, D.T., G. Faluvegi, N. Unger, E. Aguilar, G.A. Schmidt, D. Koch, S.E. Bauer, and R.L. Miller, 2006: Simulations of preindustrial, present-day, and 2100 conditions in the NASA GISS composition and climate model G-PUCCINI. *Atmospheric Chemistry and Physics*, **6(12)**, 4427-4459.

Shindell, D.T., G. Faluvegi, S.E. Bauer, D.M. Koch, N. Unger, S. Menon, R.L. Miller, G.A. Schmidt, and D.G. Streets, 2007: Climate response to projected changes in short-lived species

under an A1B scenario from 2000-2050 in the GISS climate model. *Journal of Geophysical Research*, **112(D20)**, D20103, doi:10.1029/2007JD008753.

Stockwell, W.R., F. Kirchner, M. Kuhn, and S. Seefeld, 1997: A new mechanism for regional atmospheric chemistry modeling. *Journal of Geophysical Research*, **102(D22)**, 25847-25879.

Tie, X., G. Brasseur, L. Emmons, L. Horowitz, and D. Kinnison, 2001: Effects of aerosols on tropospheric oxidants: a global model study. *Journal of Geophysical Research*, **106(D19)**, 22931-22964

Tie, X., S. Madronich, S. Walters, D.P. Edwards, P. Ginoux, N. Mahowald, R. Zhang, C. Lou, and G. Brasseur, 2005: Assessment of the global impact of aerosols on tropospheric oxidants. *Journal of Geophysical Research*, **110(D3)**, D03204, doi:10.1029/2004JD005359.

Van der Werf, G.R., J.T. Randerson, G.J. Collatz, and L. Giglio, 2003: Carbon emissions from fires in tropical and subtropical ecosystems. *Global Change Biology*, **9(4)**, 547-562.

Wesely, M.L., 1989: Parameterization of surface resistance to gaseous dry deposition in regional-scale numerical models. *Atmospheric Environment*, **23**, 1293-1304.

Wesely, M.L. and B.B. Hicks, 1977: Some factors that affect the deposition rates of sulfur dioxide and similar gases on vegetation. *Journal of the Air Pollution Control Association*, **27**, 1110-1116.

Zender, C.S., H. Bian, and D. Newman, 2003: Mineral Dust Entrainment and Deposition (DEAD) model: description and 1990s dust climatology *Journal of Geophysical Research*, **108(D14)**, 4416, doi:10.1029/2002JD002775.

Zhang, G.J. and N.A. McFarlane, 1995: Sensitivity of climate simulations to the parameterization of cumulus convection in the Canadian Climate Centre general circulation model. *Atmosphere Ocean*, **33(3)**, 407-446.

APPENDIX D REFERENCES

Collins, W.D., P.J. Rasch, B.A. Boville, J.J. Hack, J.R. McCaa, D.L. Williamson, B.P. Briegleb, C.M. Bitz, S.-J. Lin, and M. Zhang, 2006: The formulation and atmospheric simulation of the Community Atmosphere Model: CAM3. *Journal of Climate*, **19(11)**, 2144-2161.

Delworth, T.L., A.J. Broccoli, A. Rosati, R.J. Stouffer, V. Balaji, J.A. Beesley, W.F. Cooke, K.W. Dixon, J. Dunne, K.A. Dunne, J.W. Durachta, K.L. Findell, P. Ginoux, A. Gnanadesikan, C.T.

Gordon, S.M. Griffies, R. Gudgel, M.J. Harrison, I.M. Held, R.S. Hemler, L.W. Horowitz, S.A. Klein, T.R. Knutson, P.J. Kushner, A.R. Langenhorst, H.-C. Lee, S.-J. Lin, J. Lu, S.L. Malyshev, P.C.D. Milly, V. Ramaswamy, J. Russell, M.D. Schwarzkopf, E. Shevliakova, J.J. Sirutis, M.J. Spelman, W.F. Stern, M. Winton, A.T. Wittenberg, B. Wyman, F. Zeng, and R. Zhang, 2006: GFDL's CM2 global coupled climate models. Part I: Formulation and simulation characteristics. *Journal of Climate*, **19(5)**, 643-674.

Hansen, J., M. Sato, R. Ruedy, L. Nazarenko, A. Lacis, G.A. Schmidt, G. Russell, I. Aleinov, M. Bauer, S. Bauer, N. Bell, B. Cairns, V. Canuto, M. Chandler, Y. Cheng, A. Del Genio, G. Fa-luvegi, E. Fleming, A. Friend, T. Hall, C. Jackman, M. Kelley, N. Kiang, D. Koch, J. Lean, J. Lerner, K. Lo, S. Menon, R. Mill-er, P. Minnis, T. Novakov, V. Oinas, Ja. Perlwitz, Ju. Perlwitz, D. Rind, A. Romanou, D. Shindell, P. Stone, S. Sun, N. Taus-nev, D. Thresher, B. Wielicki, T. Wong, M. Yao, and S. Zhang, 2005: Efficacy of climate forcings. *Journal of Geophysical Research*, **110(D18)**, D18104, doi:10.1029/2005JD005776.

Hansen, J., M. Sato, R. Ruedy, P. Kharecha, A. Lacis, R. Miller, L. Nazarenko, K. Lo, G.A. Schmidt, G. Russell, I. Aleinov, S. Bauer, E. Baum, B. Cairns, V. Canuto, M. Chandler, Y. Cheng, A. Cohen, A. Del Genio, G. Faluvegi, E. Fleming, A. Friend, T. Hall, C. Jackman, J. Jonas, M. Kelley, N.Y. Kiang, D. Koch, G. Labow, J. Lerner, S. Menon, T. Novakov, V. Oinas, Ja. Perlwitz, Ju. Perlwitz, D. Rind, A. Romanou, R. Schmunk, D. Shindell, P. Stone, S. Sun, D. Streets, N. Tausnev, D. Thresher, N. Unger, M. Yao, and S. Zhang, 2007: Dangerous human-made interfer-ence with climate: a GISS modelE study. *Atmospheric Chemis-try and Physics*, **7(9)**, 2287-2312.

Knutson, T.R., T.L. Delworth, K.W. Dixon, I.M. Held, J. Lu, V. Ramaswamy, M.D. Schwarzkopf, G. Stenchikov, and R.J. Stouffer, 2006: Assessment of twentieth-century regional sur-face temperature trends using the GFDL CM2 coupled models. *Journal of Climate*, **10(9)**, 1624-1651.

Koch, D., T. Bond, D. Streets, N. Bell, and G.R. van der Werf, 2007: Global impacts of aerosols from particular source re-gions and sectors. *Journal of Geophysical Research*, **112**, D02205, doi:10.1029/2005JD007024.

Menon, S.A., D. Del Genio, D. Koch, and G. Tselioudis, 2002: GCM simulations of the aerosol indirect effect: sensitivity to cloud parameterization and aerosol burden. *Journal of the At-mospheric Sciences*, **59**, 692-713.

Penner, J.E., J. Quaas, T. Storelvmo, T. Takemura, O. Boucher, H. Guo, A. Kirkevag, J.E. Kristjansson, and Ø. Seland, 2006: Model intercomparison of indirect aerosol effects. *Atmospheric Chemistry and Physics Discussions*, **6**, 1579-1617.

Schmidt, G.A., R. Ruedy, J.E. Hansen, I. Aleinov, N. Bell, M. Bauer, S. Bauer, B. Cairns, V. Canuto, Y. Cheng, A. Del Genio, G. Faluvegi, A.D. Friend, T.M. Hall, Y. Hu, M. Kelley, N.Y. Kiang, D. Koch, A.A. Lacis, J. Lerner, K.K. Lo, R.L. Miller, L. Nazarenko, V. Oinas, J. Perlwitz, J. Perlwitz, D. Rind, A. Romanou, G.L. Russell, M. Sato, D.T. Shindell, P.H. Stone, S. Sun, N. Tausnev, D. Thresher, and M.-S. Yao, 2006: Present day atmospheric simulations using GISS ModelE: comparison to in-situ, satellite and reanalysis data. *Journal of Climate*, **19(2)**, 153-192.

Stouffer, R.J., T.L. Delworth, K.W. Dixon, R. Gudgel, I. Held, R. Hemler, T. Knutson, M.D. Schwarzkopf, M.J. Spelman, M.W. Winton, A.J. Broccoli, H-C. Lee, F. Zeng, and B. Soden, 2006: GFDL's CM2 global coupled climate models. Part IV: Ideal-ized climate response. *Journal of Climate*, **19(5)**, 723-740.

Wittenberg, A.T., A. Rosati, N-C. Lau, and J.J. Ploshay, 2006: GFDL's CM2 Global Coupled Climate Models. Part III: Tropical Pacific climate and ENSO. *Journal of Climate*, **19(5)**, 698-722.

APPENDIX E REFERENCES

Clarke, L., J. Edmonds, H. Jacoby, H. Pitcher, J. Reilly, and R. Richels, 2007: *Scenarios of Greenhouse Gas Emissions and Atmospheric Concentrations*. Sub-report 2.1A of Synthesis and Assessment Product 2.1 by the U.S. Climate Change Science Program and the Subcommittee on Global Change Research. Department of Energy, Office of Biological & Environmental Research, Washington, DC, 154 pp.

Nakićenović, N. and R. Swart (eds.), 2000: *Special Report on Emissions Scenarios*. A special report of Working Group III of the Intergovernmental Panel on Climate Change. Cambridge University Press, Cambridge, UK, and New York, 599 pp.

PHOTOGRAPHY CREDITS

Cover/Title Page/Table of Contents
Cover, table of contents, page 61, Image for chapter 4, (Smoke plume) Grant Goodge, STG Inc., Asheville, N.C.

Executive Summary
Page 5, (Hanging smog) Grant Goodge, STG Inc., Asheville, N.C.

Chapter 2
Page 18, (Aerial blue clouds, haze) Grant Goodge, STG Inc., Asheville, N.C.

Chapter 3

Page 29, (Golden fog in valley) Grant Goodge, STG Inc.,
Asheville, N.C.

Page 35, (Mountains between clouds) Grant Goodge, STG Inc.,
Asheville, N.C.

Page 54, (Aerial view smog over highway) Grant Goodge, STG
Inc., Asheville, N.C.

Page 57, (Philippines traffic) Deborah Misch, STG Inc., Asheville,
N.C.

Page 58, (Satellite image) Jeff Schmaltz, NOAA/Modis.

Contact Information

Global Change Research Information Office
c/o Climate Change Science Program Office
1717 Pennsylvania Avenue, NW
Suite 250
Washington, DC 20006
202-223-6262 (voice)
202-223-3065 (fax)

The Climate Change Science Program
incorporates the U.S. Global Change Research
Program and the Climate Change Research
Initiative.

To obtain a copy of this document, place
an order at the Global Change Research
Information Office (GCRIO) web site:
http://www.gcrio.org/orders

Climate Change Science Program and the Subcommittee on Global Change Research

William Brennan, Chair
Department of Commerce
National Oceanic and Atmospheric Administration
Acting Director, Climate Change Science Program

Jack Kaye, Vice Chair
National Aeronautics and Space Administration

Allen Dearry
Department of Health and Human Services

Jerry Elwood
Department of Energy

Mary Glackin
National Oceanic and Atmospheric Administration

Patricia Gruber
Department of Defense

William Hohenstein
Department of Agriculture

Linda Lawson
Department of Transportation

Mark Myers
U.S. Geological Survey

Timothy Killeen
National Science Foundation

Patrick Neale
Smithsonian Institution

Jacqueline Schafer
U.S. Agency for International Development

Joel Scheraga
Environmental Protection Agency

Harlan Watson
Department of State

EXECUTIVE OFFICE AND OTHER LIAISONS

Stephen Eule
Department of Energy
Director, Climate Change Technology Program

Katharine Gebbie
National Institute of Standards & Technology

Stuart Levenbach
Office of Management and Budget

Margaret McCalla
Office of the Federal Coordinator for Meteorology

Rob Rainey
Council on Environmental Quality

Daniel Walker
Office of Science and Technology Policy

www.ingramcontent.com/pod-product-compliance
Lightning Source LLC
Chambersburg PA
CBHW080642180526
45168CB00008B/3270